QUANTUM GRAVITY

Edited by **Rodrigo Sobreiro**

Quantum Gravity
Edited by Rodrigo Sobreiro

Published by InTech
Janeza Trdine 9, 51000 Rijeka, Croatia

Publishing Process Manager Daria Nahtigal

Technical Editor Teodora Smiljanic

Cover Designer InTech Design Team

Additional hard copies can be obtained from orders@intechopen.com

Quantum Gravity, Edited by Rodrigo Sobreiro
p. cm.
ISBN 978-953-51-0089-8

For EU product safety concerns:
IN TECH d.o.o., Prolaz Marije Krucifikse Kozulić 3, 51000 Rijeka, Croatia,
info@intechopen.com or visit our website at intechopen.com.

Meet the editor

Rodrigo F. Sobreiro graduated from Rio de Janeiro State University in 2002 and obtained his PhD from the same University in 2007. He has worked as post-doctoral researcher at Centro Brasileiro de Pesquisas Físicas and as visiting professor at Rio de Janeiro State University. Since 2009 he has a permanent professor position at Fluminense Federal University. His research interest are in quantum, classical and geometrical aspects of gauge theories and their applications in confinement and quantum gravity.

Contents

Preface

The fundamental Physics of the 20th century was constructed basically from two main theories, general relativity and quantum theory. The later allowed the construction of the standard model which describes three of the four known fundamental interactions in Nature, the exception being the gravity interaction. Unfortunately, general relativity and quantum theory have not been unified into a single coherent description of gravity in the microscopic level yet. The gravity quantization problem exists for almost one century and the final answer is yet unknown. Quantum gravity can then be considered as one of the major problems in fundamental Physics.

Many techniques to deal with the quantum nature of gravity have been developed during the last century. For instance, we can name the following techniques, popular among physicists: Canonical quantization, higher derivative gravity, Palatini-Cartan formalisms, gauge theories of the Einstein-Cartan type, Metric-affine gravities and, perhaps the most popular, the string theory. In addition to the above mentioned techniques, one can also refer to loop quantum gravity, spin foam quantum gravity, Euclidean quantum gravity and Horava-Lifshitz gravity as emergent gravity models that may describe quantum gravity. All of those techniques resulted in a massive production of interesting scientific texts.

This book presents a series of selected chapters written by renowned authors with the objective to provide an overview and comparison of the various quantum gravity theories. Each chapter treats gravity and its quantization through known and alternative techniques. The book also describes the mathematical models that have provided a framework for the themes here presented. Due to the authors efforts on writing clear yet concise texts, the reader will find each chapter very elucidative. Moreover, all contributions on this book consist on new relevant results on quantum gravity, which makes *Quantum Gravity* an useful book not only for beginners, but also for specialists of the field.

Finally, I wish to express my acknowledgements to the authors for their kindness on attending my requests regarding the format of each chapter and for the time spent to produce an important contribution to the book *Quantum Gravity*. I am also thankful to Ms. Daria Nahtigal for her patience with me and dedication to the book.

February 2012

Rodrigo F. Sobreiro
Fluminense Federal University
Instituto de Fisica, Campus da Praia Vermelha,
Avenida General Milton Tavares de Souza s/n, 24210-346,
Niteroi, RJ,
Brasil

1

Anomalous Gravitational Vacuum Fluctuations Which Act as Virtual Oscillating Dipoles

Giovanni Modanese
[1]*University of Bolzano*
[2]*Inst. for Advanced Research in the Space, Propulsion & Energy Sciences*
Madison, AL
[1]*Italy*
[2]*USA*

1. Introduction

In this work we would like to review some concepts developed over the last few years: that the gravitational vacuum has, even at scales much larger than the Planck length, a peculiar structure, with anomalously strong and long-lasting fluctuations called "zero-modes"; and that these vacuum fluctuations are virtual particles of negative mass and interact with each other, leading to the formation of weakly bound states. The bound states make up a continuum, allowing at each point of spacetime the local excitation of the gravitational vacuum through the coupling with matter in a coherent state. The spontaneous or stimulated decay of the excited states leads to the emission of virtual gravitons with spin 1 and large p/E ratio. The main results on the zero-modes and their properties have been given in (Modanese, 2011), but in this work we expand and discuss in physical terms several important details concerning the zero-mode interactions, the dynamics of virtual particles with negative mass and the properties of virtual gravitons.

Technically, our approach is based on the Lorenzian path integral of Einstein gravity in the usual metric formulation. We take the view that any fundamental theory of gravity has the Einstein action as its effective low-energy limit (Burgess, 2004). The technical problem of the non-renormalizability of the Einstein action is solved in effective quantum gravity through the asymptotic safety scheme (Niedermaier & Reuter, 2006; Percacci, 2009). According to this method, gravity can be nonperturbatively renormalizable and predictive if there exists a nontrivial renormalization group fixed point at which the infinite ultraviolet cutoff limit can be taken. All investigations carried out so far point in the direction that a fixed point with the desired properties indeed exists.

An important feature of the path integral approach is that it allows a clear visualization of the metric as a dynamical quantum variable, of which one can study averages and fluctuations also at the non-perturbative level. It is hard, however, to go much further than formal manipulations in the Lorenzian path integral; after proving the existence of the zero-modes we resort to semi-classical limits and standard perturbation theory. This method is clearly not always straightforward. At several points we proceed, by necessity, through physical induction and analogies with other interactions.

The outline of the work is the following. In Section 2 we show the existence of the zero-modes and discuss their main features, using their classical equation and the path integral. This Section contains some definitely mathematical parts, but we have made an effort to translate all the concepts in physical terms along the way. Section 3 is about the pair interactions of zero-modes: symmetric and antisymmetric states, transitions between these states, virtual dipole emission and its A and B coefficients. Section 3.3 contains a digression on the elementary dynamics of virtual particles with negative mass. Section 4 is devoted to the interaction of the zero-modes with a time-variable Λ-term. We discuss in detail the motivations behind the introduction of such a term and compare its effect to that of "regular" incoherent matter by evaluating their respective transition rates. Finally, in Section 5 we discuss in a simplified way the properties of virtual gravitons; the virtual gravitons exchanged in a quasi-static interaction are compared to virtual particles exchanged in a scattering process and to virtual gravitons emitted in the decay of an excited zero-mode.

2. Isolated zero-modes: Non trivial static metrics with null action

Our starting point is a very general property of Einstein gravity: it has a non-positive-definite action density. As a consequence, some non trivial static field configurations (metrics) exist, which have zero action. We call these configurations zero-modes of the action. The Einstein action is $S_E = -\frac{c^4}{8\pi G}\int d^4x\sqrt{g}R$ (plus boundary term; see Sect. 3) and the zero-mode condition is

$$\int d^4x\sqrt{g}R = 0 \tag{1}$$

This condition is, of course, satisfied by any metric with $R(x)$=0 everywhere (vacuum solutions of the Einstein equations (28), like for instance gravitational waves). But since the density $\sqrt{g}R$ is not positive-definite, the condition can also be satisfied by metrics which do not have $R(x)$=0 everywhere, but regions of positive and negative scalar curvature. The non-positivity of the Einstein action has been studied by Hawking, Greensite, Mazur and Mottola and others (Greensite, 1992; Mazur & Mottola, 1990). Wetterich later found that also the effective action is always un-defined in sign (Wetterich, 1998).

We are interested into these zero-action configurations because, in the Feynman path integral, field configurations with the same action tend to interfere constructively and so to give a contribution to the integral distinct from the usual classical contribution of the configurations near the stationary point of the action. Let us write the Feynman path integral on the metrics $g_{\mu\nu}(x)$ as

$$I = \int d[g]\exp\left(\frac{i}{\hbar}S_E[g]\right) \tag{2}$$

Suppose there is a subspace X of metrics with constant action. The contribution to the integral from this subspace is simply

$$I_X = \exp\left(\frac{i}{\hbar}\hat{S}_E\right)\int_X d[g] = \exp\left(\frac{i}{\hbar}\hat{S}_E\right)\mu(X) \tag{3}$$

where $\hat{S}_E$ is the constant value of the action in the subspace and $\mu(X)$ its measure. The case $\hat{S}_E = 0$ is a special case of this.

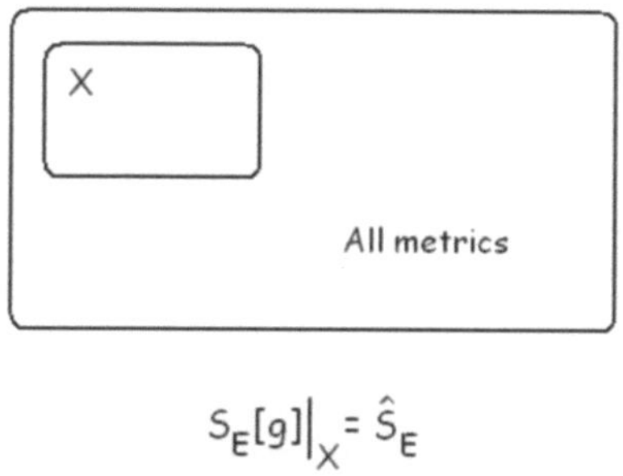

Fig. 1. Subspace X of metrics with constant action. All the metrics (spacetime configurations) in X have the same action $\hat{S}_E$. In particular, there exist a subspace whose metrics all have zero action.

The zero-modes can only give a significant contribution to the path integral if they are not isolated configurations (like a line in 2D, which has measure zero), but a whole full-dimensional subset of all the possible configurations. They are "classical" fields, not in the sense of being solutions of the Einstein equations in vacuum, but in the sense of being functions of spacetime coordinates which are weighed in the functional integral with non-vanishing measure.

2.1 Classical equation of the zero-modes

Now let us find at least some of these configurations. It is not obvious that eq. (1) has solutions with R not identically zero, because it is a difficult non-linear integro-differential equation.

In some previous work we used, to solve (1) in the weak field approximation, a method known as "virtual source method" or "reverse solution of the Einstein equations" (Modanese, 2007). According to this method, one solves the Einstein equations with non-physical sources which satisfy some suitable condition, in our case $\int dx\sqrt{g}g^{\mu\nu}T_{\mu\nu} = 0$. Since for solutions of the Einstein equations one has (trace of the equations) $R = \frac{8\pi G}{c^4}g^{\mu\nu}T_{\mu\nu}$, it follows that such solutions will be zero-modes. The expression $\int dx\sqrt{g}g^{\mu\nu}T_{\mu\nu} = 0$ is far simpler in the linear approximation. In that case the source must satisfy a condition like, for instance, $\int dxT_{00} = 0$ (supposing T_{ii} is vanishing) and is therefore a "dipolar" virtual source.

A much more interesting class of zero-modes is obtained, however, in strong field regime, starting with a spherically-symmetric Ansatz. In other words, let us look for spherically symmetric solutions of (1). Consider the most general static spherically symmetric metric

$$d\tau^2 = B(r)dt^2 - A(r)dr^2 - r^2(d\theta^2 + \sin^2\theta d\phi^2) \tag{4}$$

where $A(r)$ and $B(r)$ are arbitrary smooth functions. We add the requirement that outside a certain radius r_{ext}, $A(r)$ and $B(r)$ take the Schwarzschild form, namely

$$B(r) = \left(1 - \frac{2GM}{c^2 r}\right); \quad A(r) = \left(1 - \frac{2GM}{c^2 r}\right)^{-1} \quad \text{for } r \geq r_{ext} \tag{5}$$

This requirement serves two purposes: (1) It allows to give a physical meaning to these configurations, seen from the outside, as mass-energy fluctuations of strength M. For $r > r_{ext}$ their scalar curvature is zero. (2) More technically, the Gibbons-Hawking-York boundary term of the action is known to be constant in this case (Modanese, 2007).

Even with only the functions A and B to adjust, the condition (1) is very difficult to satisfy. We do find a set of solutions, however, if we make the drastic simplification $g_{00}=B(r)=const$. The scalar curvature multiplied by the volume element becomes in this case

$$L = \sqrt{g}R = -8\pi\sqrt{|BA|}\left(\frac{rA'}{A^2} + 1 - \frac{1}{A}\right) \tag{6}$$

Apart from the constant $c^4/8\pi G$, L is the lagrangian density of the Einstein action, computed for this particular metric. Let us fix arbitrarily a reference radius r_{ext}, and introduce reduced coordinates $s=r/r_{ext}$. Define an auxiliary function $\alpha=A^{-1}$. Regarding $L(s)$ as known, eq. (6) becomes an explicit first-order differential equation for α:

$$\alpha' = \frac{1}{s} - \frac{\alpha}{s} + \frac{L\sqrt{|\alpha|}}{8\pi s\sqrt{|B|}} \tag{7}$$

The boundary conditions (5) are written, in reduced coordinates

$$B(s \geq 1) = \left(1 - \frac{\tilde{M}}{s}\right); \quad A(s \geq 1) = \left(1 - \frac{\tilde{M}}{s}\right)^{-1} \tag{8}$$

where $\tilde{M}$ is a free parameter, the total mass in reduced units: $\tilde{M} = 2GM/c^2 r_{ext}$. In the following we shall take $\tilde{M} < 0$, in order to avoid singularities. For $r < r_{ext}$, we have $B = B(1) = 1 - \tilde{M}$.

It is interesting to note that putting $L=0$ in eq. (7) we can easily find an exact solution, ie a non-trivial static metric with $R=0$. Namely, if $1-\alpha>0$, then $\alpha=1-e^{const}/s$, which does not satisfy the boundary condition; if $1-\alpha<0$, then $\alpha=1+e^{const}/s$, implying $e^{cost} = -\tilde{M}$. The resulting g_{rr} component has the same form on the left and on the right of $s=1$, namely

$$g_{rr} = \left(1 + \frac{|\tilde{M}|}{s}\right)^{-1} \tag{9}$$

while g_{00} is constant and equal to $(1+|\tilde{M}|)$ for $s<1$, and is equal to $(1+|\tilde{M}|/s)$ for $s \geq 1$. Note that g_{rr} goes to zero at the origin.

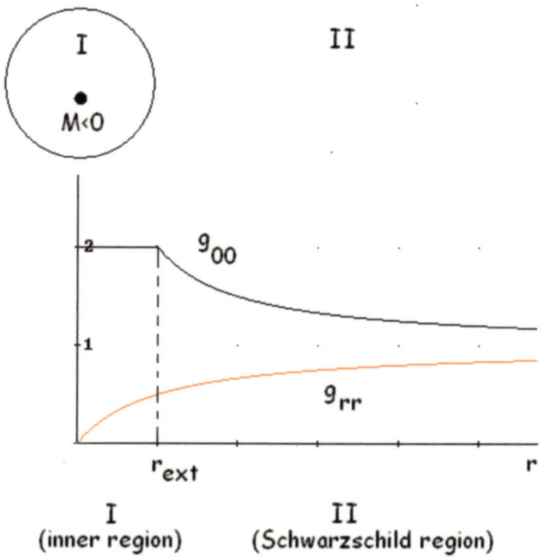

Fig. 2. Metric of an elementary static zero-mode of the Einstein action. Inside the radius r_{ext} (region I) the g_{00} component is constant, and the g_{rr} component goes to zero. On the outside (region II) both components have the form of a Schwarzschild solution with negative mass.

Now we can look for metrics close to (9), but with scalar curvature not identically zero. For large $\tilde{M}$ and small L, the last term in eq. (7) is a small perturbation. Since α never diverges and α^{-1} does not appear in the equation, the perturbed solution is not very different from (7). For values of $\tilde{M}$ of order 1 or smaller, the equation can be integrated numerically. If we choose a function $L(s)$ with null integral on the interval (0,1), we obtain a metric which is a zero-mode of the action but not of the lagrangian density. One can take, for instance, $L(s)=L_0\sin(2\pi ns)$, with n integer.

In conclusion, we have found a family of regular metrics with null scalar curvature, depending on a continuous parameter $\tilde{M}$. Furthermore, we have built a set of metrics close to the latter, by solving eq. (7) with L arbitrary but having null integral. These metrics do not have zero scalar curvature, but still have null action. They make up a full-dimensional subset of the functional space (see proof in (Modanese, 2007)).

Our solutions of the zero-mode condition are, outside the radius r_{ext}, Schwarzschild metrics with $M<0$. The quantity Mc^2 coincides with the ADM energy of the metrics. At the origin of the coordinates the component g_{rr} goes to zero, the integral of $\sqrt{g}R$ is finite and also the volume $\int dx\sqrt{g}$ is finite. The volume inside the radius r_{ext} is smaller than the volume of a sphere with the same radius in flat space.

According to our previous argument on the functional integral, these metrics give a significant contribution to the quantum averages, although they are neither classical solutions nor quantum fluctuations near the classical solutions. In the vacuum state, there exists a finite probability that the metric at any given point is not flat, but has the form of a zero-mode, i.e., seen from a distance, of a pseudo-particle of negative mass. In the language of Quantum Field Theory this could be called a vacuum fluctuation. Vacuum fluctuations are created spontaneously and at zero energetic cost at any point of spacetime, in a homogeneous and isotropic way. Usually vacuum fluctuations have a very short life, as can be shown through the Schroedinger equation (time-energy uncertainty principle) or through a transformation to Euclidean time (when the action is positive-definite). These arguments on the lifetime of the fluctuations can not be applied here, because quantum gravity has neither a local Hamiltonian, nor a positive-definite action. Our fluctuations, if they were completely isolated, would be independent of time; in fact, their interaction causes a finite lifetime (Sect. 2.3). In Sect. 5 we shall give a comparison between this kind of vacuum fluctuations and other fluctuations present in quantum gravity, like the virtual gravitons which transmit the gravitational interactions.

In order to avoid a large global curvature, the *total* average effect of the virtual masses of the zero-modes must inevitably be renormalized to zero. This is, in our view, guaranteed by the "cosmological constant paradox": nature appears to be endowed with a dynamical mechanism which relaxes to zero any constant positive or negative contributions to the vacuum energy density, coming from particle physics or even from gravity itself. So, even though such contributions are formally infinite, in the end they do not affect the curvature of spacetime. The full explanation of this mechanism can only be achieved within a complete non-perturbative theory of Quantum Gravity. Some partial evidence of the dynamical emergence of flat spacetime has been obtained in the lattice theory, and in effective field theory approaches (Hamber, 2004, Dolgov, 1997).

Therefore we shall not be concerned with the global effect of our massive vacuum fluctuations on spacetime. We shall instead consider their interactions, which result in a novel pattern of purely gravitational excited states, above a ground state in which all fluctuations pairs with equal mass are in a symmetrical superposition. Freely speaking, it's a bit like studying the local effects of pressure variations, without worrying about how the total force due to atmospheric pressure affects the Earth.

2.2 Zero-modes in the explicit functional integral

The zero-modes equation (plus the argument of non-interference) tell us that relevant run-away configurations of vacuum exist, in which the metric is locally very different from its classical value .We shall now consider an explicit path integral of Einstein gravitation, in order to evaluate the functional average of certain metric components and confirm this supposition.

Let us choose a spherical coordinate system. We integrate only over a sector X of the functional space, namely over the spherically-symmetric metric configurations with constant g_{00}. If we obtain a null quadratic vacuum average in X, namely

$$\langle g_{rr}(0)\rangle_X = \frac{\int_X d[g]\exp\left\{\frac{i}{\hbar}S[g]\right\}g_{rr}(0)}{\int_X d[g]\exp\left\{\frac{i}{\hbar}S[g]\right\}} = 0 \qquad (10)$$

this allows us to reach our conclusion: at any point there is a finite probability for a zero-mode to occur.

For these metrics the Einstein action is written (Sect. 2.1)

$$S_E = -\frac{c^4}{8\pi G}\int d^4x\sqrt{g(x)}R(x) = \frac{4\pi c^4}{G}\int dt \int_0^\infty dr\sqrt{|BA|}\left(\frac{rA'}{A^2}+1-\frac{1}{A}\right) \tag{11}$$

$$g_{\mu\nu}(x) = \eta_{\mu\nu}$$

where $A = g_{rr}$ and $B = g_{00}$ are functions of *r*. Define a radius r_{ext}, the "external radius" of our configurations, on which we impose boundary conditions as in Sect. 2.1. This means that we integrate over configurations which outside the radius r_{ext} appear like Schwarzschild metrics with mass *M*. In order to avoid singularities, we suppose $M < 0$. We can re-write the action as an integral on *r* with upper limit r_{ext}, because the scalar curvature of the Schwarzschild metric is zero. We can also add the Gibbs-Hawking-York boundary term, which in this case takes the form $S_{HGY} = -M\int dt$. For a fixed time interval, we can regard the integral $\int dt$ as a constant.

Supposing *B* constant ($B = 1 - \tilde{M}$), the path integral over these field modes is written

$$\begin{aligned}&\int d[A]\exp\left\{\frac{i}{\hbar}(S_E + S_{HGY})\right\} = \\ &= \int d[A]\exp\left\{\frac{i}{\hbar}\frac{4\pi\sqrt{|B|}}{G}\int dt\int_0^1 ds\sqrt{|A|}\left(\frac{sA'}{A^2}+1-\frac{1}{A}\right)\right\}\exp\left\{-\frac{i}{\hbar}M\int dt\right\}\end{aligned} \tag{12}$$

The second exponential can be disregarded in the functional averages, because it cancels with the normalization factor in the denominator. In the first exponential, let us define a constant factor $\alpha = \frac{1}{\hbar}\frac{4\pi\sqrt{|B|}}{G}\int dt$ and discretize the integral in *ds*. We divide the integration interval [0,1] in $(N+1)$ small intervals of length δ and replace the integral with a sum, where the derivative is written as a finite variation. We obtain

$$\int d[A]e^{\frac{i}{\hbar}S_E} = \int\prod_{j=0}^{N+1} dA_i \exp\left\{i\alpha\delta\sum_{j=0}^{N}\sqrt{|A_j|}\left(\frac{j\delta\left(A_{j+1}-A_j\right)}{\delta A_j^2}+1-\frac{1}{A_j}\right)\right\} \tag{13}$$

The presence of the square root and of the fractions with A_j makes the integrals very complicated. Let us change variables. Suppose $A > 0$, which is physically a widely justified assumption (and remember we are looking for a sufficient condition, i.e. we want to show that there exist a set of gravitational configurations for which the functional average of a quadratic quantity is different from the classical value). Define $\gamma = 1/\sqrt{A}$. This gives the new path integral

$$\int_0^{+\infty}\prod_{j=0}^{N+1} d\gamma_j \frac{2}{\gamma_j^3}\exp\left\{-i\alpha\delta\sum_{j=0}^{N}\left(2j\delta\frac{\left(\gamma_{j+1}-\gamma_j\right)}{\delta}-\frac{1}{\gamma_j}+\gamma_j\right)\right\} \tag{14}$$

(Note that $(\gamma_j + \gamma_{j+1}) \approx 2\gamma_j$ in the continuum limit.) We want to use this to compute the average $\langle \gamma_m^2 \rangle$, where m is a fixed intermediate index. This is the average of the squared field γ^2 at the point $s = m\delta$, therefore in the continuum limit it gives the average of γ^2 at the origin. We know that the system has zero-modes for which $A \to 0$ at the origin, and therefore $\gamma \to \infty$. So we would like to show that $\langle \gamma_m^2 \rangle \to \infty$ for $\delta \to 0$. This can indeed be done (Modanese, 2011), and implies in turn that (10) is true. One can also check that this is not an artefact of the continuum limit.

2.3 Zero-modes as quantum states

The explicit calculation of the average $\langle g_{rr}(0) \rangle_X$ in a sector of the functional integral is conceptually important, but in practice it does not help much in giving a quantum representation of the zero-modes and their interactions. The properties of the zero-modes as "classical" metrics are more useful for that purpose. We shall suppose that each zero-mode corresponds to a quantum state $|i\rangle$ and that $\langle i|H|i\rangle = c^2 M_i$ (see below for the meaning of the gravitational Hamiltonian H in this context). The states $|i\rangle$ are localized and mutually orthogonal. Different $|i\rangle$ correspond to field configurations centered at different points. In the following we shall also suppose for simplicity that their Schwarzschild radii are always much smaller than their distance.

According to this line of thought, the "true non-interacting ground state" of the gravitational vacuum is obtained in principle as the limit of an infinite incoherent superposition of flat spacetime (Fock vacuum) plus single zero-mode wavefunctions:

$$|0\rangle = |0\rangle_{Fock} + \sum_i \xi_i |i\rangle \tag{15}$$

This definition of the ground state is clearly difficult to put on a rigorous basis. We are mainly interested, however, into the *excitations* with respect to this ground state. The most relevant among these excitations are those resulting from pair interactions of zero-modes, as we shall see.

Note that fixing $\langle i|H|i\rangle$ amounts to a much weaker statement than giving a gravitational quantum Hamiltonian operator H, because $\langle i|H|i\rangle$ is only a matrix element and a classical limit of the total energy for an asymptotically flat configuration (ADM energy (Murchada & York, 1974)). So whenever we write here the full gravitational Hamiltonian H, in fact we only exploit some properties of its matrix elements, like in a Heisenberg representation of quantum mechanics. This is consistent with our path integral approach to the full-interacting case.

In other words, in the following we use neither the "full" gravitational Hamiltonian operator H, nor eigenvalue relations. (Interaction Hamiltonians on a background metric like that employed in Sect. 4 do not suffer from these limitations.) In fact, the Hamiltonian H is very difficult to define in quantum gravity. Even classically, there exists no generally accepted expression for the gravitational energy density. Furthermore, assuming the validity of eigenvalue operator relations would lead to contradictions. For instance, by applying the full Hamiltonian to the vacuum state (15) and supposing for a moment that $H|i\rangle = M_i c^2 |i\rangle$, we would obtain, *only formally*

$$" H|0\rangle = \sum_i \xi_i M_i c^2 |i\rangle " \tag{16}$$

From this we would conclude that $|0\rangle$ is not an eigenstate! Nevertheless the property $\langle 0|H|0\rangle = 0$ is true, considering that the coefficients ξ_i have random phases.

We could call the states $|i\rangle$ "purely gravitational, long-lived virtual particles". They are long-lived in the following sense. The classical equation for isolated zero-modes gives configurations independent from time. Adding to the pure Einstein action the boundary Gibbs-Hawking-York term, the latter takes the form $S_{GHY} = -M\int dt$, i.e. it is a constant for any fixed time interval, and does not cause interference in the path integral. However, when the zero-modes are not isolated but interact with each other, the boundary term causes their lifetime to be finite.

In the next section we shall discuss the simplest interaction of the zero-modes (pair interaction). This displays one of the typical amazing features of virtual particles (compare Sect. 5): they are created from the vacuum "for free", but after that they follow the usual dynamical rules. When computing the amplitude of a process involving virtual particles, we do not need to take into account the initial amplitude for creating the particles at a given point of space and time, but we do compute (Sect.s 3 and 4) the amplitudes for their ensuing propagation and interaction.

3. Pair interactions of zero-modes

We have introduced the concept of ground state in an effective theory of Quantum Gravity as given by the Fock vacuum plus a random superposition of zero-modes. In this Section we show that non-interacting zero-modes with equal mass are coupled in degenerate symmetric and anti-symmetric wavefunctions. The introduction of interaction removes the degeneration. The excited states form a continuum and the interaction of the vacuum with an external coherent oscillating source leads to transitions, with a probability which we shall compute in Sect. 4. As in Sect. 2, we denote with a capital M a zero-mode mass (virtual and negative).

3.1 Pairs in symmetric and antisymmetric states

Consider a couple of states $|1\rangle$ and $|2\rangle$ with masses M_1 and M_2. We have

$$\langle 1|\mathrm{H}|1\rangle = \mathrm{c}^2\mathrm{M}_1, \quad \langle 2|\mathrm{H}|2\rangle = \mathrm{c}^2\mathrm{M}_2, \quad \langle 1|2\rangle = 0 \tag{17}$$

Putting now M_1=M_2=M and taking the interaction into account, the degenerate non-interacting levels are splitted. Define the symmetrical and anti-symmetrical superpositions ψ^+ and ψ^-:

$$|\psi^+\rangle = \frac{1}{\sqrt{2}}(|1\rangle + |2\rangle) \quad |\psi^-\rangle = \frac{1}{\sqrt{2}}(|1\rangle - |2\rangle) \tag{18}$$

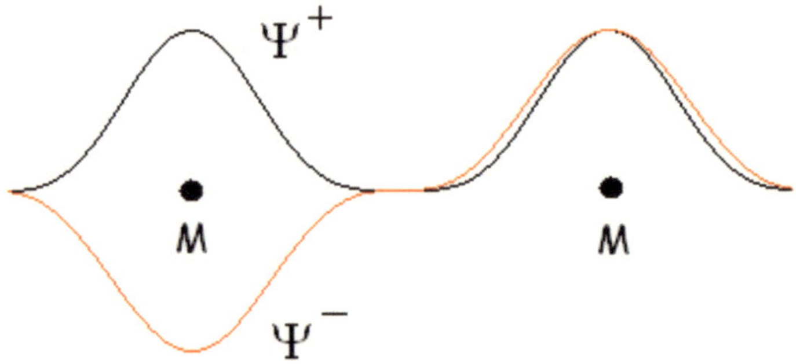

Fig. 3. Symmetric and antisymmetric bound states of zero-modes with equal mass M. (We assume that the wavefunction is much more localized near the masses than depicted – compared to their distance.)

The energy splitting ΔE is given, as known, by

$$\Delta E = E_- - E_+ = \langle \psi^- | H | \psi^- \rangle - \langle \psi^+ | H | \psi^+ \rangle = -2\langle 1 | H | 2 \rangle \tag{19}$$

Note that the matrix element $\langle 1|H|2\rangle$ can be taken to be real without loss of generality. Suppose that $\langle 1|H|2\rangle$ can be computed to a first approximation through its classical limit. The ADM energy integral at spatial infinity for the Schwarzschild-like field of two positive masses can be analytically continued to negative masses. We then obtain

$$\Delta E = 2\frac{GM^2}{r} \tag{20}$$

being r the distance between the symmetry centers of the states $|1\rangle$ and $|2\rangle$. This procedure reminds the computation of the bound states of two atoms in a molecule: the "internal states" of the atoms are not relevant and each atom is described by a single vector coordinate; the relevant Hamiltonian is the interaction Hamiltonian, although the full Hamiltonian of the system comprises in principle the forces inside the atoms and even inside the nuclei.

Let us consider the transitions between ψ^+ and ψ^-. We shall see that they are mainly of two types: (a) excitation $\psi^+ \to \psi^-$ due to the interaction with a local Λ-term dependent on time (variable vacuum energy density, associated with coherent matter - compare Sect. 4); (b) decay $\psi^- \to \psi^+$ with emission of a virtual graviton. We look for a relation between the frequency of the transition and the virtual mass of the excited states. In the ground state, all couples with equal mass will be in their symmetric superposition state. Any transition of one couple from its symmetric to its antisymmetric state gives an excited state with energy (20). Since there exist zero-modes with any (negative) mass, at any distance, there is actually a continuum of excited states.

For the same energy, in principle, there are transitions to excited levels involving different masses at different distances, provided the ratio M^2/r is the same. In practice, however, there is an upper limit on the scale r, because the time-variable Λ-term has a typical spatial extension (coherence range) of the order of 10^{-9} m, and typical frequency 10^{6}-10^{9} Hz. This fixes the maximum virtual mass involved, by eq. (20), to $M \approx 10^{-13}$ kg. This is small, but definitely much larger than any atomic scale mass, and implies that also the gravitational interaction in the pairs of virtual masses is much larger than the usual gravitational interactions at atomic scale.

We are confronted here with a very unusual situation and we should check that our description is consistent, at least at the energy scale we are considering. (In principle the zero-mode fluctuations exist at any scale, but since they are an emergent phenomenon, computed in an effective theory, it is fair to concentrate on the scale which we deem most realistic.) First, one can easily check that the supposed localization of the zero-modes is well compatible with the Heisenberg position-momentum uncertainty principle. Second, one can prove that their interaction, though strong on the atomic scale, is much weaker than the interaction in a hypothetical gravitational bound state formed by two masses of this size. This can be easily checked, for instance, by computing the corresponding Bohr radius: this is of the order of $\hbar^2 / Gm^3 \approx 10^{-19}$ m, while the zero-modes in the states ψ^+ and ψ^- are separated by a distance of the order of 10^{-9} m. So the acceleration of each zero-mode due to the presence of the other is very small, if compared to accelerations due to atomic or molecular forces. It follows that in these "weakly bound states of heavy quasi-particles" the distance r varies slowly and there is plenty of time for the transitions $\psi^+ \leftrightarrow \psi^-$ to occur at frequency 10^{6}-10^{9} Hz, as we shall describe in detail later.

On a longer time scale, the interaction itself causes the zero-modes to fade out slowly as vacuum fluctuations. This is a subtle point that completes our analysis of the isolated zero-modes given in Sect. 2. As we have seen, the boundary term $M\int dt$ in the action is constant for an isolated zero-mode, for any time interval, and therefore an isolated zero-mode will persist indefinitely in time. For interacting zero-modes the situation is more complicated, because

1. The superposition of their metrics is not necessarily a zero-mode.
2. Their total ADM mass-energy is still constant, as long as radiation is negligible; this total mass-energy comprises their masses plus potential and kinetic energy. But when the emitted radiation becomes a sizeable fraction of the total mass, the ensuing change in the boundary term in the action of the zero-modes begins to cause a destructive interference in the functional integral between the metrics $g_{\mu\nu}(x,t_1)$, $g_{\mu\nu}(x,t_2)$... at subsequent times. So the quantum amplitudes of these metrics tend to vanish and the result is that the zero-modes, as vacuum fluctuations, acquire a finite lifetime as they begin to emit dipolar or quadrupolar radiation.

3.2 Virtual dipole emission, *A* and *B* coefficients

In this Section we compute the lifetime of an excited state ψ^-. The decay of the excited state occurs with the emission of an off-shell graviton with spin 1. This happens because the dominant graviton emission process in the decay of an excited zero-mode is oscillating-dipole emission. Quadrupolar emission, which is the only process ensuring conservation of energy, momentum and spin in the emission of on-shell gravitons, can in this case be disregarded. Since we are only interested into a lowest-order perturbative estimate (tree diagrams) we can use the linearized Einstein theory in the form of the "Maxwell-Einstein" equations

$$\begin{aligned}
\nabla\cdot\mathbf{E}_G &= -4\pi G\rho_m \\
\nabla\times\mathbf{E}_G &= -\frac{\partial\mathbf{B}_G}{\partial t} \\
\nabla\cdot\mathbf{B}_G &= 0 \\
\nabla\times\mathbf{B}_G &= -\frac{4\pi G}{c^2}\mathbf{j}_m + \frac{1}{c^2}\frac{\partial\mathbf{E}_G}{\partial t}
\end{aligned} \tag{21}$$

Here $\mathbf{E}_G$ is the gravito-electric (Newtonian) field, $\mathbf{B}_G$ is the gravito-magnetic field, and $\mathbf{j}_m$, ρ_m are the mass-energy current and density. The elementary quantization of the field modes in a finite volume V follows the familiar scheme used for the computation of spontaneous and stimulated electromagnetic emission of atoms in a cavity. We have discussed in (Modanese, 2011) the conditions for applicability of the Einstein-Maxwell equations to plane waves in vacuum.

The Einstein A-coefficient of spontaneous emission turns out to be related to the B-coefficient and to the mass dipole moment by the relation

$$A = \left(B\frac{8\pi\hbar}{\lambda^3}\right) = \left(\frac{G}{\hbar^2}|\langle\hat{\mathbf{d}}\rangle|^2\frac{8\pi\hbar}{\lambda^3}\right) \tag{22}$$

where the electromagnetic coupling constants have been replaced, up to an irrelevant adimensional factor, by the gravitational constants, according to eq.s (21). The operator $\hat{\mathbf{d}}$ is the mass-dipole moment and the matrix element is taken between the initial and final state of interest.

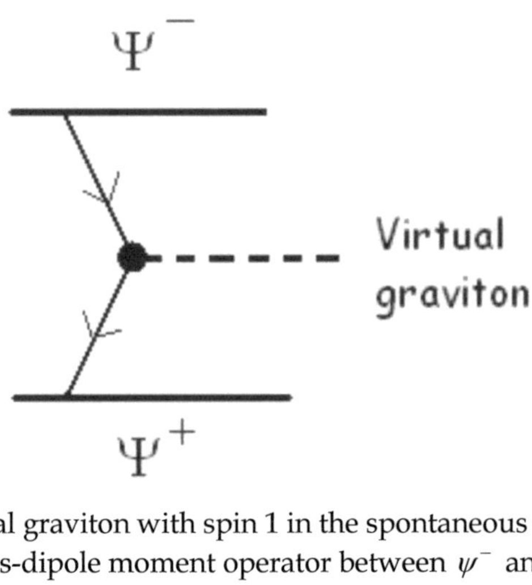

Fig. 4. Emission of a virtual graviton with spin 1 in the spontaneous decay $\psi^- \rightarrow \psi^+$. The matrix element of the mass-dipole moment operator between ψ^- and ψ^+ has module $Mr/2$.

It is straightforward to check that there is an oscillating mass dipole between the states ψ^- and ψ^+:

$$\begin{aligned}\langle\psi^+|\hat{\mathbf{d}}|\psi^-\rangle &= \frac{1}{\sqrt{2}}(\langle 1|+\langle 2|)\,\hat{\mathbf{d}}\,\frac{1}{\sqrt{2}}(|1\rangle-|2\rangle) = \\ &= \frac{1}{2}(\langle 1|+\langle 2|)(M_1\mathbf{r}_1|1\rangle - M_2\mathbf{r}_2|2\rangle) = \frac{1}{2}M\mathbf{r}\end{aligned} \tag{23}$$

where $\mathbf{r}_1 = \frac{1}{2}\mathbf{r}$, $\mathbf{r}_2 = -\frac{1}{2}\mathbf{r}$; here $\mathbf{r}$ is the displacement between the masses M_1 and M_2, which in the end are taken to be equal. The origin of the coordinate system is in the center of mass.

This mass dipole moment has purely quantum origin, because in our system there are no masses of different signs, and it is known that in this case the classical mass dipole moment computed with respect to the center of mass is zero. We could say that the non-zero matrix element (23) is due to the quantum tunnelling between the states and $|2\rangle$. This corresponds to a mass oscillation.

Eq. (22) gives the lifetime τ of the excited level ψ^- by spontaneous emission. With the values of M and r found in Section 3.1 supposing an excitation frequency of the order of 1 MHz, one finds $B \approx 10^{12}$ m³/Js² for the stimulated emission coefficient and $\tau = A^{-1} \approx 1$ s for the lifetime for spontaneous emission (taking $\lambda f \approx 1\,\mathrm{m/s}$: compare discussion in (Modanese, 2011) and Sect. 5). The general dependence of B on the frequency ω and on the length r of the dipoles is easily obtained from eq.s (20), (22) and (23):

$$B \approx \frac{1}{\hbar}\omega r^3 \tag{24}$$

Note that B is independent from the Newton constant G.

3.3 Digression: Elementary dynamics of virtual particles with negative mass

Real particles with negative mass cannot exist, because they would make the world terribly unstable, popping up spontaneously from the vacuum with production of energy. In this work, however, we hypothesize the existence of long-lived virtual particles with negative mass, whose creation from the vacuum does not require or generate any energy. We recognize that these virtual particles have negative mass by looking at their metric at infinity, which is Schwarzschild-like, but with negative M and negative ADM energy. We know that the dynamics of virtual particles, after their creation, is similar to that of real particles, and we have computed quantum amplitudes involving them.

We do not know any general principle about the "classical" dynamics of virtual particles with negative mass. Actually, virtual particles of this kind are an emergent phenomenon guessed from the path integral and can only be observed in a very indirect way. It is interesting, nonetheless, to make some reasonable hypothesis and check the consequences. Our basic assumption will be the following: for an isolated system comprising particles with positive and negative mass, the position of the center of mass, defined by

$$\mathbf{r}_{CM} = \sum_i M_i \mathbf{r}_i \tag{25}$$

is invariant in time. From this assumption one can prove in a straightforward way several strange properties of particles with negative mass. These properties can be summarized by saying that in the usual dynamical rules their mass really behaves like a negative number, namely: (a) The acceleration of the virtual particle is opposite to the applied force. (b) The momentum is opposite to the velocity. (c) The kinetic energy is negative. The kinetic energy is defined as usual through the work of the applied force, in such a way that the sum $E_{kin}+E_{pot}$ is conserved.

Applying these rules one obtains a bizarre behaviour in the scattering processes and in the dynamics. For instance, although the gravitational potential energy of two virtual particles with negative mass is negative, $E_{pot} = -GM_1M_2/r$ (compare Sect. 3.1), the two particles experience a repulsion, due to Property (a). They tend to run away from each other; while their distance increases, their E_{pot} decreases in absolute value, and their (negative) E_{kin} increases in absolute value. If the particles were initially at rest at some distance r_0 (Fig. 5), when their distance goes to infinity they gain a E_{kin} equal to their initial E_{pot}.

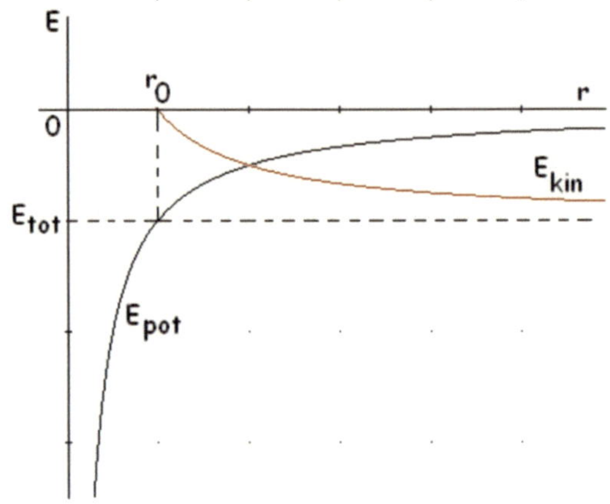

Fig. 5. "Classical" motion of two virtual particles with negative mass initially at rest at distance r_0. Although their potential energy is negative, they feel a repulsion and their (negative) kinetic energy increases in absolute value as their distance goes to infinity.

In the decay $\psi^- \to \psi^+$ (Sect. 3.2) the momentum of the emitted graviton is balanced by the recoil of the zero-modes (in the same direction of the emission). The conservation equations give

$$\begin{cases} Mv_r^2 + E_g = \Delta E \\ 2Mv_r - p_g = 0 \end{cases} \tag{26}$$

where ΔE is the energy gap, E_g and p_g are the graviton energy and momentum, v_r is the recoil velocity of the zero-mode and $2M \approx -10^{-13}$ kg is the zero-mode mass. After replacing $p_g = \alpha E_g$, the system (26) leads to the equation

$$\frac{1}{M\alpha^2} E_g^2 + E_g - \Delta E = 0 \tag{27}$$

which has a positive solution $E_g \approx \Delta E$, independently from α. Furthermore, the recoil velocity v_r turns out to be always non-relativistic. This means that the recoil of the zero-modes can always ensure conservation of momentum, independently from the value of the graviton energy-momentum ratio α.

4. Interaction of the zero-modes with a variable Λ-term

In Sect. 3 we have computed the probability of the decay process $\psi^- \to \psi^+$ with emission of a virtual graviton. The excitation process $\psi^+ \to \psi^-$ (transition of a zero-modes pair from a symmetric to an anti-symmetric state) can occur by absorption of a virtual graviton or by coupling to an external source. It is easy to show (Sect. 4.3) that the coupling of zero-modes to "ordinary" matter with energy-momentum $T_{\mu\nu} \propto m \frac{dx_\mu}{d\tau}\frac{dx_\nu}{d\tau}$ is exceedingly weak.

(Note that certain interactions between zero-modes and massive particles vanish exactly for symmetry reasons. For instance, a particle in uniform motion can never "lose energy in collisions with the zero-modes", because in its rest reference system the particle will see the vacuum, zero-modes included, as homogeneous and isotropic. There are possible exceptions to this argument: accelerated particles, or particles in states with large p uncertainty.)

The coupling to a $\Lambda(t)$ term, or local time-dependent vacuum energy density, can lead to a significant transition probability. This is due to the presence of the non-linear $\sqrt{g}$ factor in the coupling, and corresponds physically to the fact that such a Λ term does not describe isolated particles, but coherent, delocalized matter.

4.1 Summary of conventions and of some previous results

The Einstein equations with a cosmological constant, or vacuum energy term, are written

$$R_{\mu\nu} - \frac{1}{2} g_{\mu\nu} R + \Lambda g_{\mu\nu} = -\frac{8\pi G}{c^4} T_{\mu\nu} \tag{28}$$

The corresponding action (without the boundary term) is

$$S_E = -\frac{c^4}{8\pi G}\int d^4x \sqrt{g} R + \frac{\Lambda c^4}{8\pi G}\int d^4x \sqrt{g} \tag{29}$$

In this paper with use metric signature (+,-,-,-). With this convention, the cosmological (repulsive) background experimentally observed is of the order of $\Lambda c^4/G$=-10^{-9} J/m^3.

In perturbative quantum gravity on a flat background, this value of Λ corresponds to a small real graviton mass (Datta *et al.*, 2003, and ref.s). Actually, in the presence of a curved background the flat space quantization must be replaced by a suitable curved-space quantization (Novello & Neves, 2003). The limit $m \to 0$ of a theory with massive gravitons is tricky, so this global value of Λ still represents a challenge for quantum gravity (besides the need to explain its origin; compare Sect. 2.1).

In our previous work we introduced the idea that at the *local* level, the coupling of gravity with certain coherent condensed-matter systems could give an effective local positive contribution to the cosmological constant and therefore generate instabilities in the field (imaginary graviton mass (Modanese, 1996)). This early argument was not very compelling, but was reinforced after considering the effects of the Λ-term on the weak-field dipolar fluctuations mentioned in Sect. 2.1. Still such effects were predicted to be very weak and dependent on the sign and value of the background Λ at the scale of interest. After the discovery of the strong-field zero-modes of the action, in (Modanese & Junker, 2007) we computed the effect of a Λ-term on such configurations, but it still turned out to be very small.

4.2 Time-dependent Λ and zero-modes transitions

A substantial progress was made in (Modanese, 2011), where we showed that the effect of a high-frequency $\Lambda(t)$-term can be quite large and independent from its sign and from the background Λ. This was obtained considering the *interactions* between the zero-modes, as we detail in the following. Our latest computations also allow us to recognize more clearly the difference between the gravitational effect of coherent matter mediated by the Λ-term and the (negligible) gravitational effect of the classical $T_{\mu\nu}$ of the same matter. After writing the total Lagrangian $L_{grav}+L_{matter}$, we split L_{matter} into an "incoherent particles" part (Sect. 4.3) and a coherent matter part, described by a scalar field ϕ. Only the latter part contains a nonlinear factor $\sqrt{g}$, which can have non-vanishing matrix elements already to first order in Λ.

We suppose that the scalar field ϕ which describes the coherent matter has in flat space an action of the standard form

$$S_\phi = \int dx L_\phi = \int dx \left(\frac{1}{2}\partial^\mu \phi \partial_\mu \phi - \frac{1}{2} m_\phi^2 \phi^2 + k\phi^4 \right) \tag{30}$$

The gravitational coupling introduces a $\sqrt{g}$ volume factor. The dynamics of ϕ is driven by external forces, so this coupling can be regarded as an external perturbation H_Λ, a local vacuum energy density term due to the presence of coherent matter described by a macroscopic wavefunction equivalent to a classical field:

$$H_\Lambda(t,\mathbf{x}) = \frac{1}{8\pi G}\sqrt{g(t,\mathbf{x})}\Lambda(t,\mathbf{x}) = \frac{1}{8\pi G}\sqrt{g(t,\mathbf{x})}L_\phi(t,\mathbf{x}) \tag{31}$$

The Λ term in coherent matter turns out to be much larger than the cosmological background: for instance, one typically has $\Lambda c^4/G = 10^6$-10^8 J/m^3 in superconductors, depending on the type, while the currently accepted value for the cosmological background is of the order of $\Lambda c^4/G = 10^{-9}$ J/m^3. The value above for superconductors is the result of a complex evaluation of the relativistic limit of the Ginzburg-Landau Lagrangian, which yields the following expression for Λ in terms of the pairs density ρ (Modanese, 2003):

$$\Lambda(t,\mathbf{x}) = -\frac{1}{2m}\left[\hbar^2(\nabla\rho)^2 + \hbar^2\rho\nabla^2\rho - m\beta\rho^4\right] \tag{32}$$

where β is the second Ginzburg-Landau coefficient and m is the Cooper pair mass. This energy density has strong variations in space and time, following the behaviour of the macroscopic wavefunction. In order to obtain high-frequency oscillations in time, one can induce in the material Josephson currents (Modanese & Junker, 2007). Spatial variations have a typical scale of 1 nm, so we take this as the size of the volume V where the perturbation is spatially constant and the transition probability is computed. For this reason we shall leave only the time dependence in Λ and write henceforth

$$H_\Lambda(t,\mathbf{x}) = \frac{1}{8\pi G}\sqrt{g(t,\mathbf{x})}\Lambda(t) \tag{33}$$

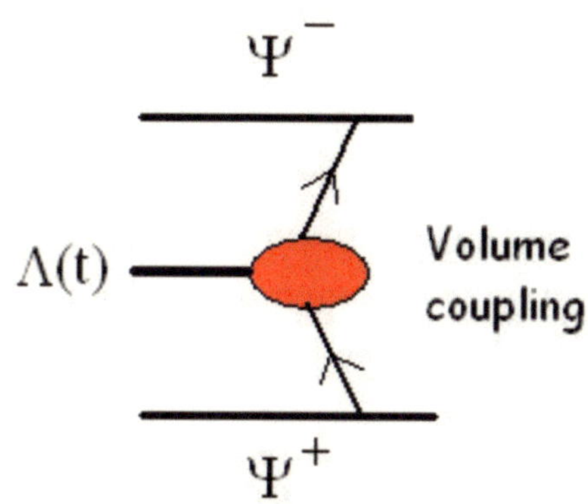

Fig. 6. A time-dependent Λ-term can be quite efficient in exciting transitions $\psi^+\to\psi^-$, because it enters the matrix elements to first order in Λ. The denomination "Volume coupling" refers to its mathematical form and to the fact that it is due to de-localized coherent matter described by a macroscopic wavefunction.

For the evaluation of the density of final states we refer to (Modanese, 2011) and give here only the final result on the probability of transitions $\psi^+\to\psi^-$ under the action of an external perturbation with oscillation frequency ω in resonance with the energy difference (20). Given the large number of available states, the resonance occurs for any frequency, and also if the perturbation is not exactly monochromatic.

In accordance with the Fermi rule and considering a volume $V\approx 10^{-27}$ m^3 and a frequency $\omega\approx 10^7$ Hz (compare Sect. 3), we obtain

$$\frac{dP}{dt} = \frac{1}{\hbar}\left|\left\langle\psi^+\middle|H_\Lambda\middle|\psi^-\right\rangle\right|^2 \rho(E) \approx 10^{34}10^{-38}10^{27} \approx 10^{23}\mathrm{s}^{-1} \tag{34}$$

This implies that the excitation time of the zero-modes in the presence of a suitable local Λ-term is very short (10^{-23} s). It is likely, actually, that this excitation process is limited by the energetic balance rather than by the transition probability.

4.3 Comparison with the effect of incoherent matter

The action of free incoherent particles is

$$S = \sum_a m_a \int \sqrt{g_{\mu\nu}(x_a)dx_a^\mu dx_a^\nu} \tag{35}$$

The index "a" denotes the sum over particles and will be omitted in the following, considering for simplicity one single particle. The corresponding field/particle interaction Hamiltonian density is

$$H_{\mathbf{x},particle} = \frac{1}{2m} h_{ij}(t,\mathbf{x}) p^i p^j \tag{36}$$

where m is the particle mass, $h_{ij} = g_{ij} - \eta_{ij}$ and i,j are spatial indices. This holds to lowest order in p and for fields h which describe a plane wave (on-shell or off-shell, see proof in (Modanese, 2011)).

Suppose to apply eq. (36) to our case, i.e. to compute a transition probability $\psi^+ \to \psi^-$ due to the coupling of gravitation to single particles in ordinary matter. In this case, the particle momentum is a given numerical function of time, while $h_{ij}(t,\mathbf{x})$ is a quantum operator which acts on the Fock vacuum creating or destroying a graviton. (In the following we shall often denote the field operator as $\hat{h}$ and omit the indices.) The numerical factor $\frac{p^i p^j}{2m}$ is of the order of the kinetic energy of the particle $\frac{p^2}{2m}$, i.e. of the order of 10^{-19} J for an atomic system. This is also the magnitude order of the $\Lambda(t)$ term. But while the interaction Hamiltonian $H_{\mathbf{x},\Lambda}$ can have non-vanishing matrix elements also when acting linearly between the states ψ^+ and ψ^-, because it is proportional to the non-linear function $\sqrt{g} = 1 + \mathrm{Tr}(h) + \ldots$, the Hamiltonian $H_{\mathbf{x},particle}$ has non-vanishing matrix elements only to second order.

Namely, we can write a matrix element of the form $\langle In|\hat{h}|Out\rangle$ as

$$\langle In|\hat{h}|Out\rangle = \langle 0, In_{z-m}|\hat{h}|0, Out_{z-m}\rangle \tag{37}$$

where $|In_{z-m}\rangle$ and $|Out_{z-m}\rangle$ denote the zero-mode components of the initial and final states, and $|0\rangle$ denotes the Fock vacuum, without gravitons. The matrix element is clearly zero, because it contains a single field acting between two Fock vacuum states. In other words, we can say that since neither in the initial state nor in the final state there are gravitons, the standard vertex (36) can have non-zero matrix element only when it is taken twice (Fig. 7) and is therefore proportional to $\left(p^2/m\right)^2$; but this is of magnitude order 10^{-38} in S.I. units, as seen, and gives a factor 10^{-76} after insertion in the transition probability (34).

On the other hand, the $\hat{h}\hat{h}$, $\hat{h}\hat{h}\hat{h}$, ... terms in the expansion of H_Λ can give non-zero matrix elements already to first order in Λ. We are not able to compute these matrix elements without a complete theory, because inside the Schwarzschild radius of the zero-modes the weak field expansion is not valid. The situation resembles that of early nuclear physics, where the nuclear matrix elements were largely unknown, apart from some general properties or magnitude orders; this did not prevent researchers from obtaining important data on the processes, based on the available information and on the crucial knowledge of the final states density.

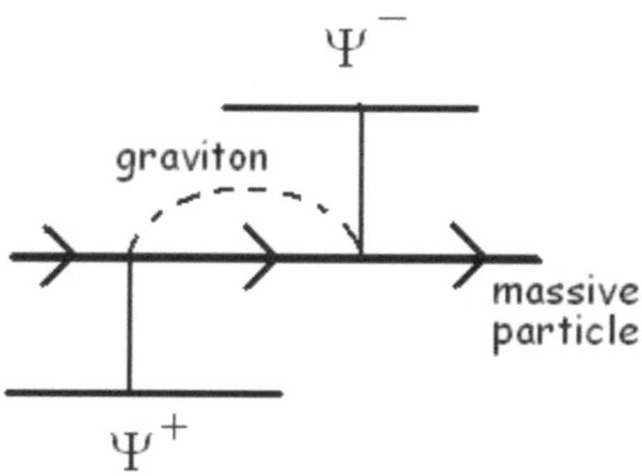

Fig. 7. Zero-mode excitation by double interaction with incoherent matter. A single massive particle can cause such an excitation by emitting and re-absorbing a virtual graviton, but the probability of this process is very small.

5. Properties of virtual gravitons

The aim of this final section is to give a simplified yet consistent physical picture of how virtual gravitons mediate the gravitational interactions. This is necessary in order to understand the link between virtual gravitons and the other kind of vacuum fluctuations studied in this paper, the zero-modes.

Note that virtual gravitons respect the usual time-energy uncertainty principle; their are not "long-lived" vacuum fluctuations like the zero-modes. This is because we consider gravitons as the particles obtained in the perturbative quantization of gravity on a flat background. It is known that the theory is not renormalizable at higher orders, but we use only tree diagrams in this work and suppose that the renormalization problem will be solved or is already solved in an effective quantum field theory of gravity (compare Sect. 1).

The concept of virtual particles mediating an interaction is not simple, and it is sometimes used improperly. In some treatments the virtual particles are seen as purely formal representations of perturbative diagrams. Instead, it is important to understand in which sense they can be regarded as particles or not.

For a real particle of given mass m, kinematics allows to connect the three quantities E, p, v through the two relations

$$E^2 - p^2c^2 = m^2c^4 \tag{38}$$

$$E = \frac{mc^2}{\sqrt{1 - v^2 / c^2}} \tag{39}$$

Therefore when one of the tree quantities is known, we can find the other two. Note that from (38) and (39) one can prove the relation $p / E = v / c^2$, which connects E and p and (unlike (39)) also holds for v=c. So we can as well consider as basic relations between E, p, v the couple

$$E^2 - p^2c^2 = m^2c^4 \tag{40}$$

$$\frac{p}{E} = \frac{v}{c^2} \tag{41}$$

These formulas all hold when the quantities m, E, p, v are well defined, thus for particles which are either stable or have a sufficiently long lifetime. For virtual particles the situation is more vague and one finds a range of statements in the literature. For instance, there is a simple textbook argument showing that the exchange of virtual photons gives rise to a $1/r^2$ force between two charges q_1 and q_2. The argument is based on the time-energy uncertainty relation. One writes $\Delta E \Delta t \approx h$, where ΔE is the energy of the exchanged virtual photon and Δt its lifetime. Supposing that the virtual particle travels with light speed, its range is $r = c\Delta t$. Therefore if the charges q_1 and q_2 are at a distance r, the "exchanged energy" is $\Delta E \propto 1/r$ and the corresponding force will be proportional to $1/r^2$. One must add the assumption that the number of exchanged photons is also proportional to the product q_1q_2 of the charges of the interacting particles. A weak point in this argument is the identification of the exchanged energy with the potential energy of the interaction. In fact, the exchanged energy depends on the velocities of the charged particles and can even be zero for static sources or in cases like that of the protons observed in their center of mass system (Fig. 8, Sect. 5.1). Apart from this, the assumption that the virtual particle has an energy uncertainty and that it propagates with light speed looks reasonable.

5.1 Example: Scattering process

Let us consider, however, another simple example: the electromagnetic scattering of two protons with the exchange of a single virtual photon. To fix the ideas, we choose a definite energy of the two protons as seen in their center of mass system, for instance $E = 10^{-13}$ J $\approx$ 1 MeV. (Magnitude orders are important in these considerations, in order to estimate the wavelength and the number of the exchanged particles, as we shall see below in the case of gravitons.) In this reference system the exchanged energy is zero and the exchanged momentum is of the order of $\sqrt{2m_pE} \approx 10^{-20}$ kg m/s (non-relativistic approximation).

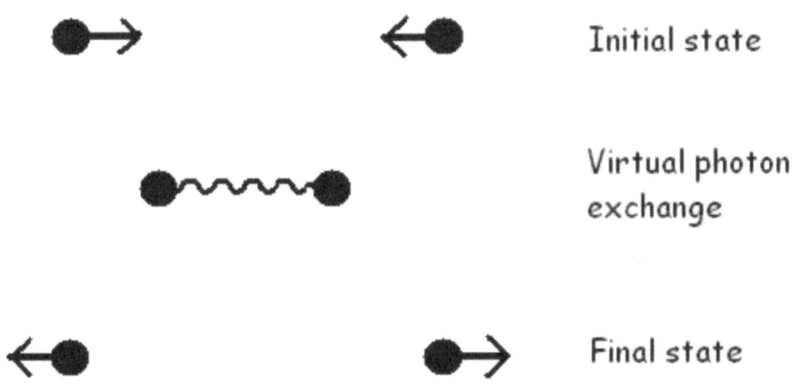

Fig. 8. Proton-proton scattering through the exchange of a single virtual photon, as seen from the center of mass reference system.

Suppose that this momentum is carried by one single virtual photon γ. The photon is off-shell, with imaginary mass $m_\gamma^2<0$: $m_\gamma^2c^2=E_\gamma^2-p_\gamma^2c^2 \Rightarrow m_\gamma^2=-p_\gamma^2/c^2$. The virtual photon energy and momentum are exactly defined and their ratio E_γ/p_γ is exactly zero in this reference system (it is not Lorentz-invariant). The wavelength of the photon, defined as $\lambda=h/p_\gamma$, is of the order of 10^{-14} m. One can estimate, classically, that the minimum distance reached by the protons is of the order of 10^{-16} m. If the virtual photon is emitted at this point, its wavefunction can clearly not be regarded as a plane wave. Its propagation velocity v is hardly observable and relation (41) appears to suggest that v is very large; if we assume $v=c$, it is only by analogy with the familiar retarded classical effects.

The situation appears, in conclusion, to be very different from the previous example. It seems reasonable to draw a clear distinction between a scattering process, which can be described as the exchange of a single virtual particle, and the inter-particle force in static or quasi-static conditions, which is in general equivalent to the exchange of a large number of virtual particles.

5.2 Photons or gravitons vs. static force

Let us now consider a different situation (Fig. 10): a massive particle (for instance, a proton) in free fall in the gravitational field of the Earth. Suppose the particle is initially at rest. There is an exact quantum formula which allows to find the static interaction potential energy in field theories like QED, QCD etc. The generalization to quantum gravity was given by (Modanese, 1995). In this formula the graviton propagator appears explicitly, as well as the G constant and the masses m_1m_2 of the sources (showing that the amplitude of virtual gravitons generation is proportional to both these masses; this property was also discussed by (Clark, 2001)). The potential energy is written as

$$U(r)=m_1m_2\lim_{T\to\infty}\frac{1}{T}\int_{-T/2}^{T/2}dt_1\int_{-T/2}^{T/2}dt_2\langle 0|h_{00}(t_1,\mathbf{r}_1)h_{00}(t_2,\mathbf{r}_2)|0\rangle \tag{42}$$

This equation describes the exchange of gravitons, for an ideally infinite time, between two static masses ($\mathbf{r}=\mathbf{r}_1-\mathbf{r}_2$; see Fig. 9). In our case the masses are the Earth and the particle.

The gravitons flux is proportional to both m_1 and m_2 and the propagator gives the amplitude of the propagation of virtual gravitons from $\mathbf{r}_1$ to $\mathbf{r}_2$, but note that their emission and absorption probabilities are equal to 1. If we expand the Feynman propagator in four-momentum space, we can see which energies and momenta are exchanged. One first finds

$$U(r)=Gm_1m_2\lim_{T\to\infty}\frac{1}{T}\int_{-T/2}^{T/2}dt_1\int_{-T/2}^{T/2}dt_2\int dE\int d^3p\frac{e^{i\mathbf{pr}-iE(t_1-t_2)}}{E^2-\vec{p}^2-i\varepsilon} \tag{43}$$

Changing variables to (t_1+t_2), (t_1-t_2) we find that the integral in (t_1+t_2) cancels the factor $1/T$. By integrating $\exp[-iE(t_1-t_2)]$ one obtains $\delta(E)$: this selects the static limit, i.e. the exchanged gravitons have $E\cong 0$ (note that in eq.s (43) and (44) we use natural units $h/2\pi=c=1$). Finally we have

$$U(r) = -Gm_1m_2 \int d^3p \frac{e^{i\mathbf{pr}}}{p^2} = -\frac{2Gm_1m_2}{\pi r} \int_0^\infty dp' \frac{\sin p'}{p'} \tag{44}$$

with $p' = |\mathbf{p}| r$. (A similar reasoning also applies to quantum electrodynamics.) The last integral is equal to $\pi / 2$ and the main contribution to the integration comes from the momentum region $p' < \pi$, i.e. $pr < \pi$. This means that in the classical interaction of two masses at distance *r*, the majority of the exchanged virtual gravitons have momentum $p < \hbar / r$ (restoring $\hbar$), or wavelength $\lambda > r$.

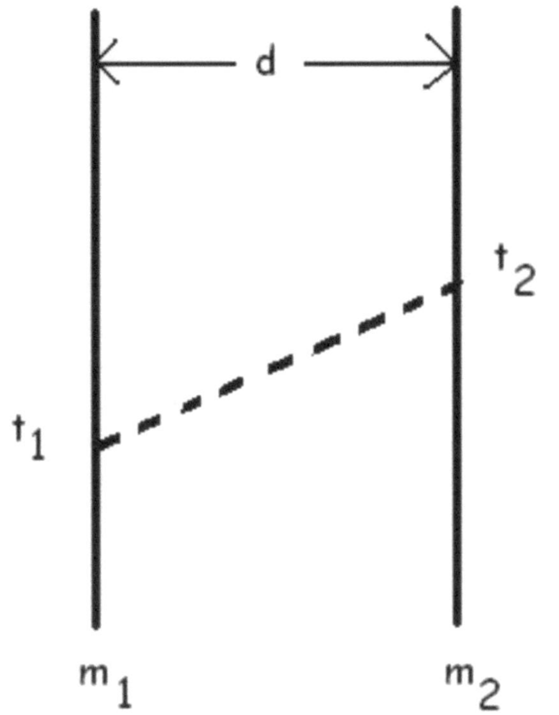

Fig. 9. Static potential energy of two masses m_1, m_2 as the outcome of graviton exchange. Virtual gravitons are emitted and absorbed at all possible times t_1, t_2; the final result is obtained by integration over t_1 and t_2.

The propagation velocity is not the same for all virtual gravitons, as is seen from the fact that their emission/absorption times vary from $-\infty$ to $+\infty$; correspondingly, their invariant masses also vary. Being a static formula, eq. (42) cannot show that the propagation velocity of the force is *c*. For this we need some generalization to moving sources; the formalism of Quantum Field Theory will ensure that the retardation effects are accounted for.

The condition $p < \hbar / r$ or $\lambda > r$ shows that the wavefunctions of the exchanged virtual gravitons are very different from plane waves: these functions do not even make a complete oscillation over a distance equal to the Earth radius! Such virtual gravitons can hardly be regarded as "particles". This should actually be expected, because the attractive character of the force can only be understood if we consider the details of the wavefunction (Baez, 1995). A naïve particle exchange can only explain repulsive forces.

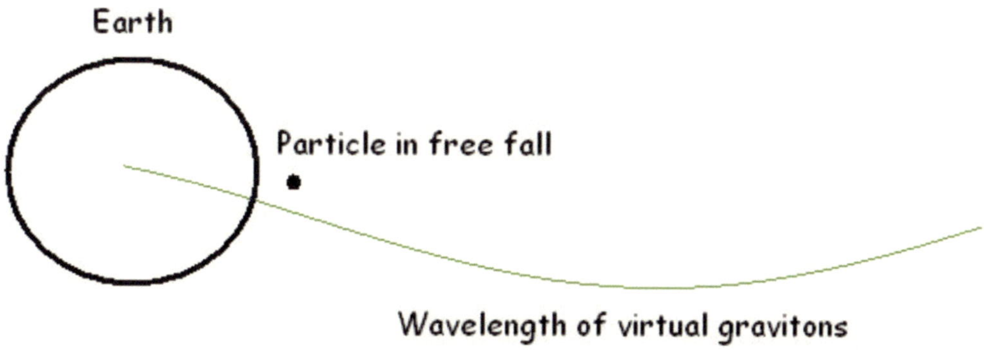

Fig. 10. The wavefunction of virtual gravitons exchanged in a quasi-static interaction is not a plane wave: for most of them, the wavelength is larger than the traveled distance.

Let us estimate how many gravitons are exchanges between the Earth and a free-falling proton or nucleon. To stay close to the static limit, consider a short interval Δt. The proton, initially at rest, acquires in this time a momentum $p=mg\Delta t$. The absorbed gravitons have, on the average, individual momentum $p \approx \hbar / r \approx 10^{-41}$ kg m/s. The number of absorbed gravitons is then of the order of $N/\Delta t \approx 10^{15}$ s^{-1}.

5.3 Virtual gravitons emitted in the decay $\psi^- \rightarrow \psi^+$

We have seen that the virtual gravitons exchanged in a quasi-static attractive gravitational interaction have very small energy and momentum. Their wavefunctions do not resemble plane waves. The propagation of this "stream" composed of a large number of virtual gravitons is a collective phenomenon occurring at light velocity.

The virtual gravitons emitted in the decay $\psi^- \rightarrow \psi^+$ (Sect. 3.2) have completely different features. Their energy is much larger ($\approx 10^{-27}$ J). Their momentum is not fixed by the emission process, since the recoil of the emitting zero-modes can balance it in any case. One of these gravitons can be individually absorbed by a real particle at rest (for instance a proton), in such a way to conserve energy and momentum, provided the product $f\lambda$ of the graviton frequency and wavelength is equal to half the final velocity of the real particle. In fact the balance equations are

$$\frac{hf}{\frac{h}{\lambda}} = \frac{E_g}{p_g} = \frac{E_p}{p_p} = \frac{\frac{1}{2}mv_p^2}{mv_p} = \frac{1}{2}v_p \tag{45}$$

where the suffix "g" denotes the virtual graviton and "p" the real particle. This is a quantum process that satisfies the conservation balance, thus it can happen and will in fact happen, with a certain amplitude. The amplitude for the final step (absorption by the real particle) is unitary, by analogy with the similar process in the static exchange.

Supposing that the real particle is a proton, it is easy to check that conservation requires $v_p \approx$ 1 m/s, $\lambda \approx 10^{-7}$ m. If the distance between the real particle and the graviton source is much larger than λ, then the wavefunction of the virtual graviton can be properly described as a plane wave. If it is legitimate to apply the kinematical relations of Sect. 5.1 to this plane wave, it follows that the virtual graviton propagates like a tachyon (Recami et al., 2000). This does not violate the causal principles of special relativity, because the propagation of a single virtual particle cannot be modulated to obtain a signal. The existence of such tachyonic virtual gravitons would be a consequence of the unique features of their source (virtual decay $\psi^- \to \psi^+$).

6. Conclusions

In quantum gravity the vacuum fluctuations have a more complex structure than in other field theories with positive-definite action. In particular, there are vacuum fluctuations which in the non-interacting approximation have infinite lifetime, and seen from the outside appear as Schwarzschild metrics with negative mass. These vacuum fluctuations behave as pseudo-particles which are created "for free" from the vacuum at any point in spacetime. The non-interacting vacuum can in fact be described as an incoherent, homogeneous and isotropic superposition of a Fock vacuum plus infinite states of this kind ("zero-modes").

When the interaction is taken into account, one finds that each pair of zero-modes with equal virtual mass M and distance r can be in two states, denoted by ψ^+ and ψ^-, with energy splitting $\Delta E = E^- - E^+ = GM^2/r$. The excited state ψ^- can decay into the state ψ^+ by emitting a virtual off-shell graviton with spin 1. The energy-momentum ratio E/p of the virtual graviton can take in principle any value, being the total momentum preserved by the recoil of the zero-modes pair. The A and B Einstein coefficients of spontaneous and stimulated emission have been computed in weak-field approximation. The B coefficient turns out to be of the order of $\omega r^2/2\pi h$, where ω is the frequency corresponding to the gap ΔE. The A coefficient depends on the wavelength; for $\omega\lambda \approx 1$ m/s one has $A \approx 1$ s^{-1}.

The excitation process $\psi^+ \to \psi^-$ cannot occur by interaction with single incoherent particles, because the relative amplitude is exceedingly small, involving a double elementary particle/graviton vertex. Instead, a sizeable excitation amplitude is obtained in the interaction with an external source of the form $\int dx \sqrt{g} \Lambda(t)$ (local vacuum energy density term, due to the presence of condensed matter in a coherent state). By taking into account the density of final states one finds, for a length scale of the Λ-term of the order of 10^{-9} m, an excitation time $\psi^+ \to \psi^-$ of the order of 10^{-23} s.

The virtual gravitons emitted in the decay $\psi^- \to \psi^+$ are very different from those exchanged in the usual gravitational interactions. Consider, for instance, a nucleon in free fall near the surface of the Earth. If it was initially at rest, it reaches a velocity of 1 m/s in approximately 0.1 s, absorbing $\approx 10^{14}$ virtual gravitons of very low frequency and large wavelength. For comparison, a single virtual graviton of frequency 10^7 Hz emitted in a vacuum decay $\psi^- \to \psi^+$ can transfer the same momentum to the nucleon in a single quick absorption process.

7. References

Burgess, C.P. (2004). Quantum Gravity in Everyday Life: General Relativity as an Effective Field Theory. *Living Reviews in Relativity*, Vol.7, p. 5

Clark, M.J. (2001). Radiation and gravitation. In: *Advances in the interplay between quantum and gravity physics*, P.G. Bergmann, V. De Sabbata (Ed.s), pp. 77-84, Kluwer, Dordrecht, The Netherlands

Datta, A. *et al.* (2003). Violation of Angular Momentum Selection Rules in Quantum Gravity. *Physics Letters B*, Vol. 579, pp. 189-199

Dolgov, A.D. (1997). Higher spin fields and the problem of cosmological constant, *Physical Review D*, Vol.55, p. 5881

Greensite, J. (1992). Vanishing of the cosmological constant in stabilized quantum gravity, Physics Letters B, Vol.291, p. 405

Hamber, HW. (2004). *Quantum Gravitation. The Feynman Path Integral Approach.* Springer, Berlin, Germany

Mazur E. & Mottola, E. (1990). The path integral measure, conformal factor problem and stability of the ground state of quantum gravity, *Nuclear Physics B*, Vol.341, pp. 187-212

McIrvin, M. (1994). Some frequently asked questions about virtual particles. Available from http://math.ucr.edu/home/baez/physics/Quantum/virtual_particles.html

Modanese, G. (1995). Potential energy in quantum gravity. *Nuclear Physics B*, Vol.434, pp. 697-708

Modanese, G. (1996). Role of a "local" cosmological constant in euclidean quantum gravity. *Physical Review D*, Vol.54, p. 5002

Modanese, G. (2003). Local contribution of a quantum condensate to the vacuum energy density. *Modern Physics Letters A*, Vol.18, pp. 683-690

Modanese, G. (2007). The vacuum state of quantum gravity contains large virtual masses. *Classical and Quantum Gravity*, Vol.24, pp. 1899-1909

Modanese, G. & Junker T. (2008). Conditions for stimulated emission in anomalous gravity-superconductors interactions. In: *Classical and Quantum Gravity Research*, M.N. Christiansen, T.K. Rasmussen, (Ed.s), 245-269, Nova Publishers, Hauppauge, USA

Modanese, G. (2011) Quantum Gravity Evaluation of Stimulated Graviton Emission in Superconductors. In: *Gravity-Superconductors Interaction: Theory and Experiment*, G. Modanese, G.A. Robertson (Ed.s), Ch. 5, Bentham, Bussum, Netherlands

Murchada N.O. & York J.W. (1974). Gravitational energy, *Physics Reviews D*, Vol.10, p. 2345

Niedermaier M. & Reuter M. (2006). The asymptotic safety scenario in quantum gravity, *Living Reviews in Relativity*, Vol.9, p. 5

Novello, M. & Neves, R.P. (2003). The mass of the graviton and the cosmological constant. *Classical and Quantum Gravity*, Vol.20, pp. L67-L73

Percacci, R. (2009). Asymptotic safety. In: *Approaches to Quantum Gravity*, D. Oriti (Ed.), Cambridge University Press, Cambridge, UK

Recami E., Fontana F., Garavaglia R. (2000). Special relativity and superluminal motions: a discussion of some recent experiments. *Int. J. Mod. Phys. A*, Vol.15, pp. 2793-2812

Wetterich, C. (1998). Effective Nonlocal Euclidean Gravity, *General Relativity and Gravitation*, Vol.30, pp. 159-172

2

Planck Scale Cosmology and Asymptotic Safety in Resummed Quantum Gravity: An Estimate of Λ

B.F.L. Ward*
Dept. of Physics,
[1]Baylor University, Waco,
[2]TH Physics Unit, CERN, Geneva
[1]USA
[2]Switzerland

1. Introduction

In Ref. [1], Weinberg suggested that the general theory of relativity may have a non-trivial UV fixed point, with a finite dimensional critical surface in the UV limit, so that it would be asymptotically safe with an S-matrix that depends on only a finite number of observable parameters. In Refs. [2–4], strong evidence has been calculated using Wilsonian [5–8] field-space exact renormalization group methods to support asymptotic safety for the Einstein-Hilbert theory. We have shown in Refs. [9–19] that the extension of the amplitude-based, exact resummation theory of Ref. [20] to the Einstein-Hilbert theory (we call the extension resummed quantum gravity) leads to UV fixed-point behavior for the dimensionless gravitational and cosmological constants, but with the bonus that the resummed theory is actually UV finite. More evidence for asymptotic safety for quantum gravity has been calculated using causal dynamical triangulated lattice methods in Ref. [21][1]. There is no known inconsistency between our analysis and Refs. [2–4, 21]. Our results are also consistent with the results on leg renormalizability of quantum gravity in Refs. [23, 24].

The reader unfamiliar with the methods of Wilson in the context of the renormalization group may consult Refs. [2, 5–8] for the details of the approach. Here we stress that in the Wilsonian formulation of the renormalization group, it does not matter whether the theory under study is actually renormalizable because the idea is to thin the degrees of freedom to those relevant to the momentum scale k under study. When one does this, the operators in the theory then fall into the classes of relevant, marginal and irrelevant operators as one studies the response of the theory to changes in the value of k. If the theory is renormalizable, then as $k \to \infty$ there will be a finite number of relevant or marginal operators in the effective action, yielding an S-matrix that depends on only a finite number of parameters. If the theory is non-renormalizable, there will be an infinite number of relevant or marginal operators in the effective action as $k \to \infty$. It was for this reason that the authors

*Work supported in part by NATO grant PST.CLG.980342.

[1] We also note that the model in Ref. [22] realizes many aspects of the effective field theory implied by the anomalous dimension of 2 at the UV-fixed point but it does so at the expense of violating Lorentz invariance.

in Ref. [2–4] have chosen to use Wilsonian methods to study the Einstein-Hilbert theory, which is naively non-renormalizable by the standard power-counting arguments. What they find is that there are only a finite number of relevant or marginal operators in the effective action as $k \to \infty$, asymptotic safety. There is no contradiction with the naive expectation because the Wilsonian methods take into the account the non-perturbative changes in the scale dimensions of the theory's operators due to interactions. Unlike the methods in Refs. [2–4] which have unphysical cut-off dependence from thinning the degrees of freedom procedures and unphysical gauge dependence, our results have no such dependence on cut-offs or gauge choice. That we agree with the findings of Refs. [2–4] then strengthens these results. Contact with experiment is now in order.

Specifically, in Ref. [25], it has been argued that the approach in Refs. [2–4] to quantum gravity may provide a realization[2] of the successful inflationary model [27, 28] of cosmology without the need of the inflaton scalar field: the attendant UV fixed point solution allows one to develop Planck scale cosmology that joins smoothly onto the standard Friedmann-Walker-Robertson classical descriptions so that one arrives at a quantum mechanical solution to the horizon, flatness, entropy and scale free spectrum problems. In Ref. [19], using the resummed quantum gravity theory [9–18], we recover the properties as used in Refs. [25] for the UV fixed point with "first principles" predictions for the fixed point values of the respective dimensionless gravitational and cosmological constants. Here, we carry the analysis one step further and arrive at a prediction for the observed cosmological constant Λ in the context of the Planck scale cosmology of Refs. [25]. We comment on the reliability of the result as well, as it will be seen already to be relatively close to the observed value [29–31]. More such reflections, as they relate to an experimentally testable union of the original ideas of Bohr and Einstein, will be taken up elsewhere [32].

The discussion is organized as follows. In the next section we review the Planck scale cosmology presented in Refs. [25]. In Section 3 we review our results [19] for the dimensionless gravitational and cosmological constants at the UV fixed point. In Section 4, we combine the Planck scale cosmology scenario [25] with our results to predict the observed value of the cosmological constant Λ. Appendix 1 contains the evaluation of our gravitational resummation exponent.

2. Planck scale cosmology

More precisely, we recall the Einstein-Hilbert theory

$$\mathcal{L}(x) = \frac{1}{2\kappa^2}\sqrt{-g}\,(R - 2\Lambda) \tag{1}$$

where R is the curvature scalar, g is the determinant of the metric of space-time $g_{\mu\nu}$, Λ is the cosmological constant and $\kappa = \sqrt{8\pi G_N}$ for Newton's constant G_N. Using the phenomenological exact renormalization group for the Wilsonian [5–8] coarse grained effective average action in field space, the authors in Ref. [25] have argued that the attendant running Newton constant $G_N(k)$ and running cosmological constant $\Lambda(k)$ approach UV fixed points as k goes to infinity in the deep Euclidean regime:

$$k^2 G_N(k) \to g_*,\ \Lambda(k) \to \lambda_* k^2$$

[2] The attendant scale choice $k \sim 1/t$ used in Refs. [25] was also proposed in Ref. [26].

for $k \to \infty$.

The contact with cosmology then proceeds as follows. Using a phenomenological connection between the momentum scale k characterizing the coarseness of the Wilsonian graininess of the average effective action and the cosmological time t, $k(t) = \frac{\xi}{t}$ for $\xi > 0$, the authors in Refs. [25] show that the standard cosmological equations admit of the following extension:

$$\begin{aligned}
(\frac{\dot{a}}{a})^2 + \frac{K}{a^2} &= \frac{1}{3}\Lambda + \frac{8\pi}{3}G_N\rho \\
\dot{\rho} + 3(1+\omega)\frac{\dot{a}}{a}\rho &= 0 \\
\dot{\Lambda} + 8\pi\rho\dot{G_N} &= 0 \\
G_N(t) &= G_N(k(t)) \\
\Lambda(t) &= \Lambda(k(t))
\end{aligned} \tag{2}$$

for the density ρ and scale factor $a(t)$ with the Robertson-Walker metric representation as

$$ds^2 = dt^2 - a(t)^2\left(\frac{dr^2}{1-Kr^2} + r^2(d\theta^2 + \sin^2\theta d\phi^2)\right) \tag{3}$$

so that $K = 0, 1, -1$ correspond respectively to flat, spherical and pseudo-spherical 3-spaces for constant time t. The equation of state is

$$p(t) = \omega\rho(t), \tag{4}$$

where p is the pressure.

Using the UV fixed points for g_* and λ_*, the authors in Refs. [25] show that the extended cosmological system given above admits, for $K = 0$, a solution in the Planck regime where $0 \leq t \leq t_{\rm class}$, with $t_{\rm class}$ a "few" times the Planck time t_{Pl}, which joins smoothly onto a solution in the classical regime, $t > t_{\rm class}$, which coincides with standard Friedmann-Robertson-Walker phenomenology but with the horizon, flatness, scale free Harrison-Zeldovich spectrum, and entropy problems all solved purely by Planck scale quantum physics. We now review the results in Refs. [19] for these UV limits as implied by resummed quantum gravity theory as presented in [9–18] and show how to use them to predict the current value of Λ. In this way, we put the arguments in Refs. [25] on a more rigorous theoretical basis.

3. g_* and λ_* in resummed quantum gravity

We start with the prediction for g_*, which we already presented in Refs. [9–19]. For the sake of completeness, let us we recapitulate the main steps in the calculation. Referring to Fig. 1, we have shown in Refs. [9–18] that the large virtual IR effects in the respective loop integrals for the scalar propagator in quantum general relativity can be resummed to the *exact* result

$$i\Delta'_F(k)|_{\rm resummed} = \frac{ie^{B''_g(k)}}{(k^2 - m^2 - \Sigma'_s + i\epsilon)} \tag{5}$$

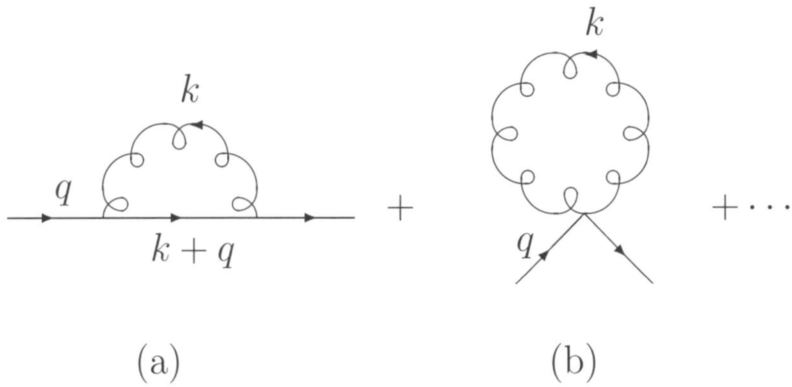

Fig. 1. Graviton loop contributions to the scalar propagator. q is the 4-momentum of the scalar.

for $(\Delta = k^2 - m^2)$

$$\begin{aligned} B''_g(k) &= -2i\kappa^2 k^4 \frac{\int d^4\ell}{16\pi^4} \frac{1}{\ell^2 - \lambda^2 + i\epsilon} \\ &\quad \frac{1}{(\ell^2 + 2\ell k + \Delta + i\epsilon)^2} \\ &= \frac{\kappa^2 |k^2|}{8\pi^2} \ln\left(\frac{m^2}{m^2 + |k^2|}\right), \end{aligned} \tag{6}$$

where the latter form holds for the UV regime, so that (5) the resummed scalar propagator falls faster than any power of $|k^2|$. An analogous result [9–18] holds for $m = 0$ (See Appendix 1.). As Σ'_s, the residual self-energy function, starts in $\mathcal{O}(\kappa^2)$, we may drop it in calculating one-loop effects. It follows that, when the respective analogs of $i\Delta'_F(k)|_{\text{resummed}}$ are used for the elementary particles, all quantum gravity loop corrections are UV finite [9–18].

We stress that our resummed scalar propagator representation (5) is not limited to the regime where $k^2 \cong m^2$ but is an identity that holds for all k^2 – see Refs. [9–18]. This is readily shown as follows. If we invert both sides of (5) and recall the formula for the exact inverse propagator, we get

$$\Delta_F^{-1}(k) - \Sigma_s(k) = (\Delta_F^{-1}(k) - \Sigma'_s(k))e^{-B''_g(k)} \tag{7}$$

where the free inverse propagator is $\Delta_F^{-1}(k) = \Delta(k) + i\epsilon$ and $\Sigma_s(k)$ is the exact proper self-energy part. We introduce here the loop expansions

$$\begin{aligned} \Sigma_s(k) &= \sum_{n=1}^{\infty} \Sigma_{s,n}(k) \left(\frac{\kappa^2}{4\pi^2}\right)^n \\ \Sigma'_s(k) &= \sum_{n=1}^{\infty} \Sigma'_{s,n}(k) \left(\frac{\kappa^2}{4\pi^2}\right)^n \end{aligned}$$

and we get, from elementary algebra, the exact relation

$$-\Sigma_{s,n}(k) = -\sum_{j=0}^{n} \Sigma'_{s,j}(k) \left(\frac{-4\pi^2 B''_g(k)}{\kappa^2}\right)^{n-j} / (n-j)! \tag{8}$$

where we define for convenience $-\Sigma_{s,0}(k) = -\Sigma'_{s,0}(k) = \Delta_F^{-1}(k)$. This proves that every Feynman diagram contribution to $\Sigma_s(k)$ corresponds to a unique contribution to $\Sigma'_s(k)$ to all orders in $\kappa^2/(4\pi)$ for all values of k^2. QED.

When we use our resummed propagator results, as extended to all the particles in the SM Lagrangian and to the graviton itself, , working now with the complete theory (see Refs. [9–19]) of (1) plus the SM Lagrangian written in diffeomorphism invariant form as explained in Refs. [9–18],

$$\mathcal{L}(x) = \frac{1}{2\kappa^2}\sqrt{-g}\,(R - 2\Lambda) + \sqrt{-g}L^{\mathcal{G}}_{SM}(x) \tag{9}$$

where $L^{\mathcal{G}}_{SM}(x)$ is the SM Lagrangian written in diffeomorphism invariant form as explained in Refs. [9–18], the denominator of the graviton propagator becomes [9–18] (M_{Pl} is the Planck mass)

$$q^2 + \Sigma^T(q^2) + i\epsilon \cong q^2 - q^4\frac{c_{2,eff}}{360\pi M_{Pl}^2}, \tag{10}$$

for

$$\begin{aligned} c_{2,eff} &= \sum_{\text{SM particles j}} n_j I_2(\lambda_c(j)) \\ &\cong 2.56 \times 10^4 \end{aligned} \tag{11}$$

with I_2 given in Refs. [9–18] by

$$I_2(\lambda_c) = \int_0^\infty dx x^3 (1+x)^{-4-\lambda_c x} \tag{12}$$

and with $\lambda_c(j) = \frac{2m_j^2}{\pi M_{Pl}^2}$. n_j is the number of effective degrees of freedom [9–18] of particle j of mass m_j. In $c_{2,eff}$ in (11), we take the SM masses as explained in Refs. [9–19] following Refs. [29–31, 33–35] : for the now presumed three massive neutrinos [33], we estimate a mass at $\sim$ 3 eV; for the remaining members of the known three generations of Dirac fermions $\{e, \mu, \tau, u, d, s, c, b, t\}$, we use [34] $m_e \cong 0.51$ MeV, $m_\mu \cong 0.106$ GeV, $m_\tau \cong 1.78$ GeV, $m_u \cong 5.1$ MeV, $m_d \cong 8.9$ MeV, $m_s \cong 0.17$ GeV, $m_c \cong 1.3$ GeV, $m_b \cong 4.5$ GeV and $m_t \cong 174$ GeV and for the massive vector bosons $W^\pm$, Z we use the masses $M_W \cong 80.4$ GeV, $M_Z \cong 91.19$ GeV, respectively. We set the Higgs mass at $m_H \cong 120$GeV, in view of the limit from LEP2 [35]. We note that (see the Appendix 1 here and the Appendix 1 in Ref. [9]) when the rest mass of particle j is zero, such as it is for the photon and the gluon, the value of m_j turns-out to be $\sqrt{2}$ times the gravitational infrared cut-off mass [29–31], which is $m_g \cong 3.1 \times 10^{-33}$eV. We also note that from Ref.[36] it also follows that the value of n_j for the graviton and its attendant ghost is 42. For $\lambda_c \to 0$, we have found the approximate representation

$$I_2(\lambda_c) \cong \ln\frac{1}{\lambda_c} - \ln\ln\frac{1}{\lambda_c} - \frac{\ln\ln\frac{1}{\lambda_c}}{\ln\frac{1}{\lambda_c} - \ln\ln\frac{1}{\lambda_c}} - \frac{11}{6}. \tag{13}$$

We thus identify (we use G_N for $G_N(0)$)

$$G_N(k) = G_N/(1 + \frac{c_{2,eff}k^2}{360\pi M_{Pl}^2}) \tag{14}$$

and compute the UV limit g_* as

$$g_* = \lim_{k^2 \to \infty} k^2 G_N(k^2) = \frac{360\pi}{c_{2,eff}} \cong 0.0442. \tag{15}$$

We stress that this result has no threshold/cut-off or gauge effects in it. a pure property of the known world.

Turning now to λ_*, we use Einstein's equation

$$G_{\mu\nu} + \Lambda g_{\mu\nu} = -\kappa^2 T_{\mu\nu} \tag{16}$$

in a standard notation where $G_{\mu\nu} = R_{\mu\nu} - \frac{1}{2} R g_{\mu\nu}$, $R_{\mu\nu}$ is the contracted Riemann tensor, and $T_{\mu\nu}$ is the energy-momentum tensor. Working with the representation $g_{\mu\nu} = \eta_{\mu\nu} + 2\kappa h_{\mu\nu}$ for the flat Minkowski metric $\eta_{\mu\nu} = \text{diag}(1, -1, -1, -1)$ we may isolate Λ in Einstein's equation by evaluating its VEV (vacuum expectation value). For any bosonic quantum field φ we use the point-splitting definition (here, : : denotes normal ordering)

$$\begin{aligned}\varphi(0)\varphi(0) &= \lim_{\epsilon \to 0} \varphi(\epsilon)\varphi(0) \\ &= \lim_{\epsilon \to 0} T(\varphi(\epsilon)\varphi(0)) \\ &= \lim_{\epsilon \to 0} \{: (\varphi(\epsilon)\varphi(0)) : + < 0|T(\varphi(\epsilon)\varphi(0))|0 >\}\end{aligned} \tag{17}$$

where the limit is taken with time-like $\epsilon \equiv (\epsilon, \vec{0}) \to (0,0,0,0) \equiv 0$ respectively. A scalar then makes the contribution [9–18] to Λ given by[3]

$$\begin{aligned}\Lambda_s &= -8\pi G_N \frac{\int d^4k}{2(2\pi)^4} \frac{(2k_0^2) e^{-\lambda_c (k^2/(2m^2)) \ln(k^2/m^2+1)}}{k^2 + m^2} \\ &\cong -8\pi G_N [\frac{1}{G_N^2 64 \rho^2}],\end{aligned} \tag{18}$$

where $\rho = \ln \frac{2}{\lambda_c}$ and we have used the calculus of Refs. [9–18]. We note that the standard equal-time (anti-)commutation relations algebra realizations then show that a Dirac fermion contributes -4 times Λ_s to Λ. The deep UV limit of Λ then becomes

$$\begin{aligned}\Lambda(k) &\underset{k^2 \to \infty}{\longrightarrow} k^2 \lambda_*, \\ \lambda_* &= -\frac{c_{2,eff}}{2880} \sum_j (-1)^{F_j} n_j / \rho_j^2 \\ &\cong 0.0817\end{aligned} \tag{19}$$

where F_j is the fermion number of j and $\rho_j = \rho(\lambda_c(m_j))$. We see again that λ_* is free of threshold/cut-off effects and of gauge artifacts and is a pure prediction of our known

[3] We note the use here in the integrand of $2k_0^2$ rather than the $2(\vec{k}^2 + m^2)$ in Ref. [19], to be consistent with $\omega = -1$ [37] for the vacuum stress-energy tensor.

world – λ_* would vanish in an exactly supersymmetric theory. Our results for (g_*, λ_*) agree qualitatively with those in Refs. [2, 25].

For reference, we note that, if we restrict our resummed quantum gravity calculations above for g_*, λ_* to the pure gravity theory with no SM matter fields, we get the results

$$g_* = .0533,\ \lambda_* = -.000189$$

. We see that our results suggest that there is still significant cut-off effects in the results used for $g_*,\ \lambda_*$ in Refs. [2, 25], which already seem to include an effective matter contribution when viewed from our resummed quantum gravity perspective, as an artifact of the obvious gauge and cut-off dependences of the results. Indeed, from a purely quantum field theoretic point of view, the cut-off action is

$$\Delta_k S(h, C, \bar{C}; \bar{g}) = \frac{1}{2} < h, \mathcal{R}_k^{\text{grav}} h > + < \bar{C}, \mathcal{R}_k^{\text{gh}} C > \tag{20}$$

where $\bar{g}$ is the general background metric, which is the Minkowski space metric η here, and $C, \bar{C}$ are the ghost fields and the operators $\mathcal{R}_k^{\text{grav}},\ \mathcal{R}_k^{\text{gh}}$ implement the course graining as they satisfy the limits

$$\lim_{p^2/k^2 \to \infty} \mathcal{R}_k = 0,$$

$$\lim_{p^2/k^2 \to 0} \mathcal{R}_k \to \mathfrak{Z}_k k^2,$$

for some $\mathfrak{Z}_k$ [2]. Here, the inner product is that defined in the second paper in Refs. [2] in its Eqs.(2.14,2.15,2.19). The result is that the modes with $p \lesssim k$ have a shift of their vacuum energy by the cut-off operator. There is therefore no disagreement in principle between our gauge invariant and cut-off independent results and the gauge dependent and cut-off dependent results in Refs. [2, 25].

4. An estimate of Λ

To estimate the value of Λ today, we take the normal-ordered form of Einstein's equation,

$$: G_{\mu\nu} : + \Lambda : g_{\mu\nu} := -\kappa^2 : T_{\mu\nu} : . \tag{21}$$

The coherent state representation of the thermal density matrix then gives the Einstein equation in the form of thermally averaged quantities with Λ given by our result above in lowest order. Taking the transition time between the Planck regime and the classical Friedmann-Robertson-Walker regime at $t_{tr} \sim 25 t_{Pl}$ from Refs. [25], we introduce

$$\begin{aligned} \rho_\Lambda(t_{tr}) &\equiv \frac{\Lambda(t_{tr})}{8\pi G_N(t_{tr})} \\ &= \frac{-M_{Pl}^4(k_{tr})}{64} \sum_j \frac{(-1)^F n_j}{\rho_j^2} \end{aligned} \tag{22}$$

and use the arguments in Refs. [38] (t_{eq} is the time of matter-radiation equality) to get the first principles estimate, from the method of the operator field,

$$\begin{aligned}
\rho_\Lambda(t_0) &\cong \frac{-M_{Pl}^4(1+c_{2,eff}k_{tr}^2/(360\pi M_{Pl}^2))^2}{64}\sum_j \frac{(-1)^F n_j}{\rho_j^2} \\
&\quad \times [\frac{t_{tr}^2}{t_{eq}^2}\times(\frac{t_{eq}^{2/3}}{t_0^{2/3}})^3] \\
&\cong \frac{-M_{Pl}^2(1.0362)^2(-9.197\times 10^{-3})}{64}\frac{(25)^2}{t_0^2} \\
&\cong (2.400\times 10^{-3}eV)^4.
\end{aligned} \tag{23}$$

where we take the age of the universe to be $t_0 \cong 13.7\times 10^9$ yrs. In the latter estimate, the first factor in the square bracket comes from the period from t_{tr} to t_{eq} (radiation dominated) and the second factor comes from the period from t_{eq} to t_0 (matter dominated) [4]. This estimate should be compared with the experimental result [29–31][5] $\rho_\Lambda(t_0)|_{\text{expt}} \cong (2.368\times 10^{-3}eV(1\pm 0.023))^4$.

To sum up, our estimate, while it is definitely encouraging, is not a precision prediction, as possible hitherto unseen degrees of freedom have not been included and t_{tr} is not precise, yet.

5. Acknowledgments

We thank Profs. L. Alvarez-Gaume and W. Hollik for the support and kind hospitality of the CERN TH Division and the Werner-Heisenberg-Institut, MPI, Munich, respectively, where a part of this work was done.

6. Appendix: Gravitational infrared exponent

In the text, we use several limits of the gravitational infrared exponent B_g'' defined in (6). This appendix contains these evaluations for completeness.

We have to consider

$$-B_g''(p) = \frac{2i\kappa^2 p^4}{16\pi^4}\int \frac{d^4k}{(k^2-\lambda^2+i\epsilon)}\frac{1}{(k^2-2kp+\Delta+i\epsilon)^2} \tag{24}$$

where $\Delta = p^2 - m^2$. The integral on the RHS of (24) is given by

$$\begin{aligned}
I &= \int \frac{d^4k}{(k^2-\lambda^2+i\epsilon)}\frac{1}{(k^2-2kp+\Delta+i\epsilon)^2} \\
&= \frac{-i\pi^2}{p^2}\frac{1}{x_+-x_-}\left[x_+\ln(1-1/(\sqrt{2}x_+)) - x_-\ln(1-1/(\sqrt{2}x_-))\right]
\end{aligned}$$

[4] The method of the operator field forces the vacuum energies to follow the same scaling as the non-vacuum excitations.

[5] See also Ref. [39] for an analysis that suggests a value for $\rho_\Lambda(t_0)$ that is qualitatively similar to this experimental result.

with

$$x_{\pm} = \frac{1}{2\sqrt{2}}\left(\bar{\Delta} + \bar{\lambda}^2 \pm ((\bar{\Delta} + \bar{\lambda}^2)^2 - 4(\bar{\lambda}^2 - i\bar{\epsilon}))^{1/2}\right) \tag{25}$$

for $\bar{\Delta} = 1 - m^2/p^2$, $\bar{\lambda}^2 = \lambda^2/p^2$and $\bar{\epsilon} = \epsilon/p^2$. In this way, we arrive at the results, for $p^2 < 0$,

$$B''_g(p) = \begin{cases} \frac{\kappa^2|p^2|}{8\pi^2}\ln\left(\frac{m^2}{m^2+|p^2|}\right), & m \neq 0 \\ \frac{\kappa^2|p^2|}{8\pi^2}\ln\left(\frac{m_g^2}{m_g^2+|p^2|}\right), & m = m_g = \lambda \\ \frac{2\kappa^2|p^2|}{8\pi^2}\ln\left(\frac{m_g^2}{|p^2|}\right), & m = 0,\ m_g = \lambda \end{cases} \tag{26}$$

where we have made more explicit the presence of the observed small mass, m_g, of the graviton. When m=0 and one wants to use dimensional regularization for the IR regime instead of m_g, we normalize the propagator at a Euclidean point $k^2 = -\mu^2$ and use standard factorization arguments [40–44] to take the factorized result for B''_g from (26) as

$$B''_g(p)|_{\text{factorized}} = \frac{2\kappa^2|p^2|}{8\pi^2}\ln\left(\frac{|\mu^2|}{|p^2|}\right), \quad m = 0,\ m_g = 0. \tag{27}$$

In physical applications, such mass singularities are absorbed by the definition of the initial state "parton" densities and/or are canceled by the KLN theorem in the final state; we do not exponentiate them in the exactly massless case.

We stress that the standard analytic properties of the 1PI 2pt functions obtain here, as we use standard Feynman rules. Wick rotation changes the Minkowski space Feynman loop integral $\int d^4k$ with $k = (k^0, k^1, k^2, k^3)$ for real k^j and $k^2 = k^{0^2} - k^{1^2} - k^{2^2} - k^{3^2}$ into the integral $i\int d^4k_E$ with $k = (ik^0, k^1, k^2, k^3)$ and $k^2 = -k^{0^2} - k^{1^2} - k^{2^2} - k^{3^2} \equiv -k_E^2$ with k_E the Euclidean 4-vector $k_E = (k^0, k^1, k^2, k^3)$ with metric $\delta_{\mu\nu} = diag(1,1,1,1)$. Thus our results rigorously correspond to $|p^2| = -p^2$ in (26), (27) with m^2 replaced with $m^2 - i\epsilon$, with $\epsilon \downarrow 0$, following Feynman, for $p^2 < 0$; by Wick rotation this is the regime relevant to the UV behavior of the Feynman loop integral. Standard complex variables theory then uniquely specifies our exponent for any value of p^2.

7. References

[1] S. Weinberg, in *General Relativity, an Einstein Centenary Survey*, eds. S. W. Hawking and W. Israel, (Cambridge Univ. Press, Cambridge, 1979).
[2] M. Reuter, Phys. Rev. D 57 (1998) 971; O. Lauscher and M. Reuter, *ibid.* 66 (2002) 025026; A. Bonanno and M. Reuter, *ibid.* 62 (2000) 043008, and references therein.
[3] D. F. Litim, Phys. Rev. Lett. 92(2004) 201301; Phys. Rev. D 64 (2001) 105007, and references therein.
[4] R. Percacci and D. Perini, *Phys. Rev. D* 68 (2003) 044018.
[5] K. G. Wilson, Phys. Rev. B 4 (1971) 3174, 3184; K. G. Wilson, J.Kogut, Phys. Rep. 12 (1974) 75.
[6] F. Wegner and A. Houghton, Phys. Rev. A 8(1973) 401.
[7] S. Weinberg, "Critical Phenomena for Field Theorists", *Erice Subnucl. Phys.* (1976) 1.
[8] J. Polchinski, Nucl. Phys. B 231 (1984) 269.
[9] B.F.L. Ward, Open Nucl.Part.Phys.Jour. 2(2009) 1.

[10] B.F.L. Ward, Mod. Phys. Lett. A 17 (2002) 237.
[11] B.F.L. Ward, Mod. Phys. Lett. A 19 (2004) 14.
[12] B.F.L. Ward, J. Cos. Astropart. Phys. 0402 (2004) 011.
[13] B.F.L. Ward, Acta Phys. Polon. B37 (2006) 1967.
[14] B.F.L. Ward, Acta Phys. Polon. B37 (2006) 347.
[15] B.F.L. Ward, hep-ph/0502104; in *Focus on Black Hole Research*, ed. P.V. Kreitler,(Nova Sci. Publ., Inc., New York, 2006) p. 95.
[16] B.F.L. Ward, hep-ph/0411050; Int. J. Mod. Phys. A20 (2005) 3502.
[17] B.F.L. Ward, hep-ph/0411049; Int. J. Mod. Phys. A20 (2005) 3128.
[18] B.F.L. Ward, hep-ph/0410273; in *Proc. ICHEP 2004*, vol. 1, eds. H. Chen *et al.*,(World Sci. Publ. Co., Singapore, 2005) p. 419 and references therein.
[19] B.F.L. Ward, Mod. Phys. Lett. A 23 (2008) 3299.
[20] D. R. Yennie, S. C. Frautschi, and H. Suura, Ann. Phys. 13 (1961) 379; see also K. T. Mahanthappa, Phys. Rev. 126 (1962) 329, for a related analysis.
[21] J. Ambjorn *et al.*, Phys. Lett. B 690 (2010) 420, and references therein.
[22] P. Horava, Phys. Rev. D 79 (2009) 084008.
[23] D. Kreimer, Ann. Phys. 321 (2006) 2757.
[24] D. Kreimer, Ann. Phys. 323 (2008) 49.
[25] A. Bonanno and M. Reuter, Phys. Rev. D 65 (2002) 043508; Jour. Phys. Conf. Ser. 140 (2008) 012008, and references therein.
[26] I.L. Shapiro and J. Sola, Phys. Lett. B 475 (2000) 236.
[27] See for example A. H. Guth and D.I. Kaiser, Science 307 (2005) 884; A. H. Guth, Phys. Rev. D 23 (1981) 347, and references therein.
[28] See for example A. Linde, Lect. Notes. Phys. 738 (2008) 1, and references therein.
[29] A.G. Riess *et al.*, Astron. Jour. 116 (1998) 1009.
[30] S. Perlmutter *et al.*, Astrophys. J. 517 (1999) 565.
[31] C. Amsler *et al.*, Phys. Lett. B 667 (2008) 1 and, references therein.
[32] B.F.L. Ward, to appear.
[33] See for example D. Wark, in *Proc. ICHEP02*, eds. S. Bentvelsen *et al.*, (North-Holland,Amsterdam, 2003), Nucl. Phys. B (Proc. Suppl.) 117 (2003) 164; M. C. Gonzalez-Garcia, hep-ph/0211054, in *Proc. ICHEP02*, eds. S. Bentvelsen *et al.*, (North-Holland,Amsterdam, 2003), Nucl. Phys. B (Proc. Suppl.) 117 (2003) 186, and references therein.
[34] K. Hagiwara *et al.*, Phys. Rev. D 66 (2002) 010001; S. Eidelman *et al.*, Phys. Lett. B 592 (2004) 1; H. Leutwyler and J. Gasser, Phys. Rept. 87 (1982) 77, and references therein.
[35] D. Abbaneo *et al.*, hep-ex/0212036; M. Gruenewald, hep-ex/0210003, in *Proc. ICHEP02*, eds. S. Bentvelsen *et al.*, (North-Holland,Amsterdam, 2003), Nucl. Phys. B Proc. Suppl. 117(2003) 280.
[36] G. 't Hooft and M. Veltman, Ann. Inst. Henri Poincare XX (1974) 69.
[37] Ya. B. Zeldovich, Sov. Phys. Uspekhi 11 (1968) 381.
[38] V. Branchina and D. Zappala, G. R. Gravit. 42 (2010) 141; arXiv:1005.3657, and references therein.
[39] J. Sola, J. Phys. A 41 (2008) 164066.
[40] R.K. Ellis *et al.*, Phys. Lett. B 78 (1978) 281-4.
[41] R.K. Ellis *et al.*, Nucl. Phys. B 152 (1979) 285-329.
[42] D. Amati, R. Petronzio and G. Veneziano G, Nucl. Phys. Bbf 146 (1978) 29-49.
[43] S. Libby and G. Sterman G, Phys. Rev. D 18 (1978) 3252-68.
[44] A. Mueller, Phys. Rev. D 18 (1978) 3705-27.

3

S-Duality in Topological Supergravity

Eckehard W. Mielke and Alí A. Rincón Maggiolo*

Departamento de Física,
Universidad Autónoma Metropolitana–Iztapalapa,
México

1. Introduction

One of the main motivations for constructing a model of topological gravity in *three dimensions* (3D) is that it might serve as a 'laboratory' for applying techniques appearing rather awkward or even intractable in four dimensions. This stems from the fact that a Riemannian spacetime is Ricci-flat, i.e., the Ricci tensor determines the Riemann tensor in 3D and as a result, the only vacuum solutions of the Einstein equations with vanishing cosmological constant are flat. This result implies that the dynamical properties may not be attributed to the metric but rather to the coframe. When matter is included there are nontrivial solutions to the Einstein equations and if topological terms are included, these may induce dynamical properties in 3D. Such a 'laboratory' may no longer be a suitable testing ground for higher–dimensional models of Einsteinian gravity [5, 10, 18, 36].

There are other reasons for studying the dynamical aspects of topological gravity in three dimensions: Some problems in 4D gravity reduce to an effective 3D theory, such as cosmic strings, the high–temperature behavior of 4D theories and some membrane models of extended relativistic systems. Moreover, many aspects of black hole thermodynamics can be effectively reduced to problems in 3D, cf. Refs. [6, 7].

Outside of quantum gravity, the continuum theory of lattice defects in crystal physics can be regarded as 'analogue gravity' with Cartan's torsion in 3D, where such defects are modeled by connections in the orthonormal frame bundle and the Chern-Simons type free-energy integral by Riemann–Cartan (RC) spaces with constant torsion [11, 26]. Recently, flexural modes of graphene have also been considered as membranes with a 'gravitational' metric [25] or coframe induced from its embedding into three-dimensional spacetime.

Our paper is organized as follow: In Section 2, we give a brief introduction to the Mielke-Baekler (MB) model of toplogical gravity in 3D, in which the Einstein-Cartan Lagrangian is substituted by a *mixed* topological term, the so-called *mix*-model. The coupling of Rarita-Schwinger fields to topological gravity is presented in Section 3, whereas in Section 4 we deduce the restrictions on the coupling parameters in order to ensure that the model is supersymmetric. The particular dynamical symmetry of the MB model, in Ref. [32] dubbed "S–duality", is generalized in Section 5 to our topological supergravity model. In Section 6 and in an Outlook, we consider the still speculative applicability of this model to the

*Permanent address: Departamento de Física, Facultad de Ciencias, Universidad del Zulia, Venezuela

	objects	p-forms	components	n=4	3	2
ϑ^α	vector	1	n^2	16	9	1
$\Gamma^\star_\alpha$	vector	1	n^2	16	9	1
T^α	vector	2	$n^2(n-1)/2$	24	9	2
$R^{\alpha\beta}$	bivector	2	$n^2(n-1)^2/4$	36	9	1
Σ_α	vector	$n-1$	n^2	16	9	4
$\tau_{\alpha\beta}$	bivector	$n-1$	$n^2(n-1)/2$	24	9	2
η_α	vector	$n-1$	n^2	16	9	4

Table 1. Geometrical objects and fields

flexural modes of corrugated surfaces (2D membranes) embedded in 3D spacetime, as recently realized by the rather prospective new material called *graphene*.

2. Topological gravity with torsion

In three spacetime dimensions, the basic gravitational variables in the Riemann-Cartan (RC) formalisms are the one–forms of the coframe and the Lie dual of the (Lorentz-) rotational connection $\Gamma^{\beta\gamma} = \Gamma_j{}^{\beta\gamma} dx^j$, i.e.,

$$\vartheta^\alpha = e_i{}^\alpha dx^i \quad \text{and} \quad \Gamma^\star_\alpha := \frac{1}{2}\eta_{\alpha\beta\gamma}\Gamma^{\beta\gamma}. \tag{1}$$

The related field strengths are the two–forms of torsion

$$T^\alpha := d\vartheta^\alpha - (-1)^s\, \eta^{\alpha\beta} \wedge \Gamma^\star_\beta \tag{2}$$

and curvature

$$R^\star_\alpha = \frac{1}{2}\eta_{\alpha\beta\gamma}R^{\beta\gamma} := d\Gamma^\star_\alpha + \frac{(-1)^s}{2}\eta_{\alpha\beta\gamma}\Gamma^\star_\beta \wedge \Gamma^\star_\gamma, \tag{3}$$

respectively, cf. the Appendices of Ref. [31]. Table 1 contains a summary of these basic variables and their components in various dimensions. Observe that only for $n = 3$ all fields have the same number of components. After converting bivectors into vectors via the Lie dual, a linear combination of all variables could pave the way for a better understanding of topological models.

In 3D, the Einstein-Cartan (EC) Lagrangian

$$L_{\rm EC} := -\frac{\chi}{\ell}\vartheta^\alpha \wedge R^\star_\alpha \equiv -\chi\, C_{\rm TL} - \frac{\chi}{\ell}\, d(\Gamma^\star_\alpha \wedge \vartheta^\alpha) \tag{4}$$

merely gives rise to a locally trivial dynamics [38]. This is due to its equivalence to a 'mixed' Chern-Simons type term $C_{\rm TL}$ plus a total divergence, as indicated above.

In this paper, we generalize this trivial dynamics by adding Chern-Simons (CS) type terms, following the lead of Witten [43]. By gauging the Poincaré group $\mathbb{R}^3 \rtimes SO(1,2)$, we arrive at the Mielke and Baekler (MB) model [2, 28] which is at most *linear* in the field strengths. This is slightly modified here by replacing $L_{\rm EC}$ via the 'mixed' Chern-Simons type term $C_{\rm TL}$, which is simulating, in 3D, to some extend Einstein's theory with 'cosmological' term, as is indicated above. Thereby, we are able to depart from a completely topological theory.

Allowing for arbitrary "vacuum angles" $\theta_{\rm T}$, $\theta_{\rm L}$ and $\theta_{\rm TL} = -\chi$, the most general purely *topological* gravity Lagrangian in 3D, in first order formalism, takes the form

$$L_{\rm MB}(\vartheta^\alpha, \Gamma^\star_\alpha) = \theta_{\rm T} C_{\rm T} + \theta_{\rm L} C_{\rm L} + \theta_{\rm TL} C_{\rm TL}\,, \tag{5}$$

where

$$C_{\rm T} := \frac{1}{2\ell^2}\vartheta^\alpha \wedge T_\alpha, \qquad C_{\rm L} := (-1)^s\,\Gamma^{\star\alpha} \wedge R^\star_\alpha - \frac{1}{3!}\,\eta_{\alpha\beta\gamma}\,\Gamma^{\star\alpha} \wedge \Gamma^{\star\beta} \wedge \Gamma^{\star\gamma} \tag{6}$$

and

$$C_{\rm TL} := \frac{1}{\ell}\left(\Gamma^{\star\alpha} \wedge T_\alpha - \frac{(-1)^s}{2}\,\eta_{\alpha\beta\gamma}\,\Gamma^{\star\alpha} \wedge \Gamma^{\star\beta} \wedge \vartheta^\gamma\right), \tag{7}$$

respectively, are the translational, rotational and 'mixed' Chern-Simons type three forms of gauge type $C = {\rm Tr}\{A \wedge F\}$ in RC spacetime [8, 12, 43]. The equation (5) is the known topological Lagrangian of the Mielke-Baekler (MB) *mix*-model [28, 31]. Since the translational term $C_{\rm T}$ is covariant, it appears that the MB model is semi-topological, with interesting consequence on the degrees of propagating modes, cf. Ref. [4, 32, 36].

Varying the Lagrangian (5) with respects to ϑ^α and $\Gamma^{\star\alpha}$ and employing the results of Appendix A, yields the topological field equations

$$-\,\theta_{\rm TL}\,R^\star_\alpha - \frac{\theta_{\rm T}}{\ell}\,T_\alpha = \ell\,\Sigma_\alpha\,, \tag{8}$$

and

$$-\,(-1)^s\theta_{\rm TL}\,T_\alpha - \frac{\theta_{\rm T}}{2\ell}\,\eta_\alpha - \theta_{\rm L}\ell\,R^\star_\alpha = \ell\,\tau^\star_\alpha\,, \tag{9}$$

cf. Eq. (6.9) of Ref. [2]. Observe that the translational CS term proportional to $\theta_{\rm T}$ induces in the second field equation a constant term, familiar in 4D from Einstein's equation with cosmological constant Λ.

Thereby, combining the *vacuum* field equations (9) and (8) yield for the torsion and the RC curvature the constrictions:

$$T_\alpha = \frac{2\kappa}{\ell}\eta_\alpha\,, \qquad R^\star_\alpha = \frac{\rho}{\ell^2}\eta_\alpha \tag{10}$$

where the contortional constants $\kappa = \theta_{\rm TL}\theta_{\rm T}/2A$ and $\rho = -\theta^2_{\rm T}/A$ are related to the vacuum angles. A singular case is exclude by assuming that $A =: -(-1)^s\theta^2_{\rm TL} + 2\theta_{\rm T}\theta_{\rm L} \neq 0$.

When including matter couplings, we explicitly find for the torsion

$$T_\alpha - \frac{2\kappa}{\ell}\eta_\alpha = \frac{2}{A}\ell\left(\theta_{\rm TL}\tau^\star_\alpha - \theta_{\rm L}\ell\Sigma_\alpha\right), \tag{11}$$

and the RC curvature

$$R^\star_\alpha - \frac{\rho}{\ell^2}\eta_\alpha = \frac{2}{A}\left(\theta_{\rm TL}\ell\Sigma_\alpha - \theta_{\rm T}\tau^\star_\alpha\right), \tag{12}$$

cf. Ref. [31].

3. Rarita–Schwinger Lagrangian in 3D

Commonly, *supergravity* [15, 19] with one supersymmetry generator, i.e. $\mathcal{N}$=1, represents the simplest consistent coupling of a Rarita–Schwinger (RS) spin-$\frac{3}{2}$ field [35] to gravity.

The Rarita-Schwinger [35] type spinor-valued one-form[1]

$$\Psi = \Psi_i dx^i = \Psi_\alpha \vartheta^\alpha \tag{13}$$

can be written holononically and anholononically. However, it does not depend on the coframe, inasmuch $\Psi_\alpha := e_\alpha \rfloor \Psi$ involves the inverse tetrad. In 3D, we adhere to the conventions that the holonomic indices run from $i, j, k, \ldots = 0, 1, 2$, whereas $\alpha, \beta, \ldots = \hat{0}, \hat{1}, \hat{2}$ for the anholonomic indices.

We are going to provide a brief summary of the spinors that will be used in three dimensions: As well known, the covering group of the rotation group $SO(3)$ is isomorphic to the unitary group $SU(2)$. Since an element of $SU(2)$ can be parameterized by three numbers, the most convenient basis of the Lie algebra are the familiar Pauli spin matrices:

$$\sigma^1 = \begin{pmatrix} 0 & 1 \\ 1 & 0 \end{pmatrix}, \quad \sigma^2 = \begin{pmatrix} 0 & -i \\ i & 0 \end{pmatrix}, \quad \sigma^3 = \begin{pmatrix} 1 & 0 \\ 0 & -1 \end{pmatrix}. \tag{14}$$

These matrices satisfy the following Lie algebra:

$$\left[\sigma^\alpha, \sigma^\beta\right] = 2i\eta^{\alpha\beta\gamma}\sigma_\gamma. \tag{15}$$

However, for Lorentzian signature $s = 1$, the covering group of $SO(1,2)$ is isomorphic to the real group $SL(2,\mathbb{R})$. Then the generators of $SL(2,\mathbb{R})$ may be realized by the matrices

$$\gamma_0 = i\sigma^2, \quad \gamma_1 = \sigma^1, \quad \gamma_2 = \sigma^3. \tag{16}$$

These real matrices [27] satisfying

$$\gamma_\alpha \gamma_\beta = g_{\alpha\beta} + \eta_{\alpha\beta\nu}\gamma^\nu \tag{17}$$

also provide a realization of the Clifford algebra

$$\gamma_\alpha \gamma_\beta + \gamma_\beta \gamma_\alpha = 2g_{\alpha\beta} \tag{18}$$

in 3D. In addition, the coframe basis ϑ^α converts into one Clifford algebra value one-form

$$\gamma = \gamma_\alpha \vartheta^\alpha \tag{19}$$

Then Ψ will become real two-component spinors, with the Dirac adjoint defined by $\overline{\Psi} := \Psi^\dagger \gamma^0$.

[1] In four dimensions (4D), the Rarita–Schwinger field $\Psi := \Psi_\alpha \vartheta^\alpha$ entering Eq. (13) is a *Majorana spinor* valued one-form. As it is well known [34], it satisfies the Majorana condition, i.e. $\Psi = C\overline{\Psi}^t$, where C is the charge conjugation matrix given by $C = -i\gamma_0$ satisfying $C^\dagger = C^{-1}, \quad C^t = -C$ and $C^{-1}\gamma^\alpha C = -(\gamma^\alpha)^t$. Consequently,

$$\overline{\Psi} \wedge \Psi = 0, \quad \overline{\Psi} \wedge \gamma_5 \gamma^\alpha \Psi = 0, \quad \overline{\Psi} \wedge \gamma_5 \Psi = 0$$

For the real Majorana representation all γ^α are purely imaginary and the components of the gravitino vector–spinor consequently are all real [30].

The corresponding *manifestly Hermitian* RS type Lagrangian three–form of Howe and Tucker [23] reads

$$L_{\rm RS} = \frac{i}{4}\left(\overline{\Psi} \wedge D\Psi - \Psi \wedge \overline{D\Psi}\right) + \frac{i}{4}\, m\overline{\Psi} \wedge \gamma \wedge \Psi, \tag{20}$$

including, however, a mass term. Here minimal coupling to gravity is achieved via

$$D\Psi = d\Psi - \frac{1}{2}\gamma_\alpha \Gamma^{\star\alpha} \wedge \Psi, \tag{21}$$

which is nothing more than the gauge covariant derivative of a spinor-valued one-form Ψ.

Only in 3D, however, there exists a generalization given by the following expression

$$L_\Psi = L_{\rm RS} + s_1\, \overline{D\Psi} \wedge {}^*(D\Psi) + s_2\, \overline{D\Psi} \wedge \gamma \wedge {}^*(\gamma \wedge D\Psi). \tag{22}$$

As in the case of the Rarita-Schwinger Lagrangian $L_{\rm RS}$, it is manifestly Hermitian when the additional quadratic derivative terms carry s_1 and s_2 as dimensionless coupling constants.

In order to supersymmetrize this action, it will be coupled to topological gravity later on.

3.1 Energy-momentum and spin currents

By definition, the energy-momentum current two-form Σ_α of matter is given by

$$\Sigma_\alpha := \frac{\delta L_\Psi}{\delta \vartheta^\alpha} = \frac{\partial L_\Psi}{\partial \vartheta_\alpha} + D\frac{\partial L_\Psi}{\partial T^\alpha}, \tag{23}$$

where the second term accounts for the possibility of a non-minimal coupling to torsion via Pauli type terms, cf. Eq. (5.1.8) of Ref. [22]. According to the Noether theorem, the energy-momentum current two-form of matter Σ_α without Pauli terms can be rewritten as

$$\Sigma_\alpha := e_\alpha \rfloor L_\Psi - (e_\alpha \rfloor \Psi) \wedge \frac{\partial L_\Psi}{\partial \Psi} - (e_\alpha \rfloor \overline{\Psi}) \wedge \frac{\partial L_\Psi}{\partial \overline{\Psi}} - (e_\alpha \rfloor D\Psi) \wedge \frac{\partial L_\Psi}{\partial D\Psi} - (e_\alpha \rfloor D\overline{\Psi}) \wedge \frac{\partial L_\Psi}{\partial D\overline{\Psi}}, \tag{24}$$

see Eq. (5.4.11) of Ref. [22] for details. This equivalent equation often is more convenient, since it involves only partial derivatives of the matter fields and avoids the intricate treatment of a possible dependence of the matter Lagrangian on the Hodge dual. Taking into account the identities of Appendix B, we find

$$\begin{aligned}\Sigma_\alpha = & -\frac{i}{4} m\overline{\Psi} \wedge \gamma_\alpha \Psi + s_1 \left\{\overline{D\Psi} \wedge e_\alpha \rfloor {}^*(D\Psi) - (e_\alpha \rfloor D\Psi) \wedge {}^*\left(\overline{D\Psi}\right)\right\} \\ & + s_2 \left[\overline{D\Psi} \wedge \gamma_\alpha \wedge {}^*(\gamma \wedge D\Psi) - (e_\alpha \rfloor D\Psi) \wedge {}^*\left(\overline{D\Psi} \wedge \gamma\right) \wedge \gamma\right].\end{aligned} \tag{25}$$

Since the kinetic terms in the Rarita-Schwinger type Lagrangian $L_{\rm RS}$ do not depend explicitly on the coframe ϑ^α, they provides no contribution to the energy-momentum current.

The 3-dual of the spin current is defined by

$$\tau^\star_\alpha := \frac{1}{2}\eta_{\alpha\beta\gamma}\tau^{\beta\gamma} = \frac{(-1)^s}{2}\frac{\delta L_\Psi}{\delta \Gamma^\star_\alpha}. \tag{26}$$

In view of the definition (21) of the covariant derivative, we find

$$\tau^{\star}_{\alpha} = \frac{(-1)^s}{2}\Big\{\frac{i}{4}\overline{\Psi}\gamma_\alpha\Psi + \frac{s_1}{2}\Big[\overline{\Psi}\gamma_\alpha\wedge{}^*(D\Psi) + \gamma_\alpha\Psi\wedge{}^*\left(\overline{D\Psi}\right)\Big]$$
$$+\frac{s_2}{2}\Big[\overline{\Psi}\gamma_\alpha\wedge\gamma\wedge{}^*(\gamma\wedge D\Psi) + \gamma_\alpha\Psi\wedge\gamma\wedge{}^*\left(\gamma\wedge \overline{D\Psi}\right)\Big]\Big\}. \tag{27}$$

Using the Hermetian properties of the spinor-valued p–forms, we finally obtain

$$\tau^{\star}_{\alpha} = \frac{(-1)^s}{2}\left[\frac{i}{4}\overline{\Psi}\wedge\gamma_\alpha\Psi + s_1\,\overline{\Psi}\gamma_\alpha\wedge{}^*(D\Psi) + s_2\,\overline{\Psi}\gamma_\alpha\wedge\gamma\wedge{}^*(\gamma\wedge D\Psi)\right], \tag{28}$$

cf. the identities of Appendix C.

It should be noted that for the pure Rarita-Schwinger Lagrangian with $s_1 = s_2 = 0$, the energy-momentum current is proportional to its dual spin, i.e.

$$\Sigma_\alpha = -(-1)^s 2m\tau^{\star}_{\alpha}\,. \tag{29}$$

4. Topological supersymmetry in 3D

Let us consider the first order topological Lagrangian

$$L_\infty = L_\infty(\vartheta^\alpha, \Gamma^{\star}_{\alpha}, \Psi) = L_{\text{MB}} + L_\Psi \tag{30}$$

and verify if it is supersymmetric or not: The variation of its independent variables $(\vartheta^\alpha, \Gamma^{\star}_{\alpha}, \Psi)$ yields

$$\delta L = \delta\vartheta^\alpha\wedge\frac{\delta L}{\delta\vartheta^\alpha} + \delta\Gamma^{\star}_{\alpha}\wedge\frac{\delta L}{\delta\Gamma^{\star}_{\alpha}} + \delta\overline{\Psi}\wedge\frac{\delta L}{\delta\overline{\Psi}} \tag{31}$$

where, for convenience, it suffices to vary only for the Dirac adjoint $\overline{\Psi}$.

The supersymmetric transformation of Deser [13, 14] read in exterior form notation

$$\delta_{\text{susy}}\vartheta^\alpha = i\overline{\sigma}\,\Psi\gamma^\alpha, \qquad \delta_{\text{susy}}\Gamma^{\star}_{\alpha} = i\overline{\sigma}\,\gamma^{*}_{\alpha}D\Psi + ic\overline{\sigma}\left(\gamma_\alpha\Psi + e_\alpha\rfloor{}^*\Psi\right),$$
$$\delta_{\text{susy}}\Psi = 2D\sigma + c\gamma\sigma, \tag{32}$$

where σ stands in for a spinor valued zero form and c a real constant. Inserting this into Eq. (31) yields

$$\delta_{\text{susy}}L = i\overline{\sigma}\,\Psi\gamma^\alpha\wedge\frac{\delta L}{\delta\vartheta^\alpha} + \delta_{\text{susy}}\Gamma^{\star}_{\alpha}\wedge\frac{\delta L}{\delta\Gamma^{\star}_{\alpha}} + \left(2\overline{D\sigma} + c\overline{\sigma}\gamma\right)\wedge\frac{\delta L}{\delta\overline{\Psi}}, \tag{33}$$

where we used $\overline{c\gamma\sigma} = c\overline{\sigma}\gamma$ for the Dirac adjoint.

In the following, we assume that the second field equation $\delta L/\delta\Gamma^{\star}_{\alpha} \cong 0$ is fulfilled "on shell", i.e., Eq. (9) of the 'mixed' MB model. Then, the SUSY transformation reduce to

$$\delta_{\text{susy}}L \cong \overline{\sigma}\left(i\gamma^\alpha\Psi\wedge\frac{\delta L}{\delta\vartheta^\alpha} - 2D\wedge\frac{\delta L}{\delta\overline{\Psi}} + c\gamma\wedge\frac{\delta L}{\delta\overline{\Psi}}\right) + 2d\left(\overline{\sigma}\wedge\frac{\delta L}{\delta\overline{\Psi}}\right) \tag{34}$$

Let us restrict for the moment to the usual Rarita-Schwinger Lagrangian $L_{\rm RS}$, or equivalently to L_Ψ with $s_1 = s_2 = 0$. Then the Rarita-Schwinger equation

$$\frac{2}{i}\frac{\delta L}{\delta \overline{\Psi}} = D\Psi + \frac{1}{2} m\gamma \wedge \Psi \cong 0 \tag{35}$$

becomes massive. Moreover, in Eq. (34) the term in brackets following form the supersymmetric transformations reads

$$\begin{aligned} & i\gamma^\alpha \Psi \wedge \frac{\delta L}{\delta \vartheta^\alpha} + c\gamma \wedge \frac{\delta L}{\delta \overline{\Psi}} - 2D\frac{\delta L}{\delta \overline{\Psi}} \\ & \quad \cong i\gamma^\alpha\Psi \left(\frac{\theta_{\rm TL}}{\ell} R^\star_\alpha + \frac{\theta_{\rm T}}{\ell^2} T_\alpha + \Sigma_\alpha \right) + c\gamma \wedge \left(\frac{i}{2} D\Psi + \frac{i}{4} m\gamma \wedge \Psi \right) \\ & \quad -D \left(iD\Psi + \frac{i}{2} m\gamma \wedge \Psi \right) \\ & \quad = i\gamma^\alpha\Psi \left(\frac{\theta_{\rm TL}}{\ell} R^\star_\alpha + \frac{\theta_{\rm T}}{\ell^2} T_\alpha \right) + \gamma^\alpha \Psi \left(\frac{1}{4} m \overline{\Psi}\gamma_\alpha \Psi \right) \\ & \quad +c\gamma \wedge \left(\frac{i}{2} D\Psi + \frac{i}{4} m\gamma \wedge \Psi \right) - iR^\star_\alpha \gamma^\alpha \Psi - \frac{i}{2} m T_\alpha \gamma^\alpha \Psi + \frac{i}{2} m\gamma \wedge D\Psi \end{aligned} \tag{36}$$

By a Fierz rearrangement, i.e.,

$$\gamma^\alpha \Psi \wedge \overline{\Psi}\gamma_\alpha \Psi = 0, \tag{37}$$

terms arising from the energy-momentum current Σ_α, or likewise from the dual spin $\tau^\star_\alpha$, are vanishing.

Moreover, in our restricted model with $s_1 = s_2 = 0$ we have to put

$$c = -m, \tag{38}$$

in order to eliminate the kinetic $\gamma \wedge D\Psi$ terms. Then, using the formula

$$\gamma \wedge \gamma = -2\gamma^\alpha \eta_\alpha, \tag{39}$$

of Howe and Tucker [23], we find from Eq. (36) the requirement

$$i \left[\left(\frac{\theta_{\rm TL}}{\ell} - 1 \right) R^\star_\alpha + \left(\frac{\theta_{\rm T}}{\ell^2} - \frac{m}{2} \right) T_\alpha + \frac{m^2}{2} \eta_\alpha \right] \gamma^\alpha \Psi = 0, \tag{40}$$

in order that our Lagrangian becomes supersymmetric.

At first sight, it appears that there is no cosmological constant in order to compensate a similar one arising from the RS mass. However, one should compare the bracket with the second field equation (9) *inserted*, which indeed involves a cosmological term induced by the translational Chern-Simons term proportional to $\theta_{\rm T}$. In this insertion

$$i \left[\left(\theta_{\rm L} + \frac{\theta_{\rm TL}}{\ell} - 1 \right) R^\star_\alpha + \left((-1)^s \frac{\theta_{\rm TL}}{\ell} + \frac{\theta_{\rm T}}{\ell^2} - \frac{m}{2} \right) T_\alpha + \frac{1}{2} \left(\frac{\theta_{\rm T}}{\ell^2} + m^2 \right) \eta_\alpha + \tau^\star_\alpha \right] \gamma^\alpha \Psi = 0, \tag{41}$$

the dual spin $\tau^{\star}_{\alpha}$ of the RS field will not contribute, again due to Fierz rearrangement (37). This finally leads to the "on shell" conditions

$$\theta_{\rm T} = -m^2\ell^2, \qquad \theta_{\rm TL} = \frac{(-1)^s}{2} m(2m+1)\ell, \qquad \theta_{\rm L} = 1 - \frac{\theta_{\rm TL}}{\ell} = 1 - \frac{(-1)^s}{2} m(2m+1) \quad (42)$$

for the coupling constants of the bosonic part of our Lagrangian L_∞. Consequently, massless RS spinors do not require a translational nor a 'mixed' CS term in order to acquire supersymmetry.

5. Towards supersymmetric S–duality

There exists a continuous deformation [or a field redefinition (FR)] of the (Lorentz-) rotational connection by adding a tensor–valued one–form, similarly as in Eq. (3.11.1) of Ref. [22]. In 3D, the particular deformation

$$\widetilde{\Gamma}^{\star}_{\alpha} = \Gamma^{\star}_{\alpha} - (-1)^s \frac{\varepsilon}{2\ell} \vartheta_\alpha, \quad (43)$$

where ε is a continuous parameter, is involving the Lie dual $\Gamma^{\star}_{\alpha} = \frac{1}{2}\eta_{\alpha\beta\gamma}\Gamma^{\beta\gamma}$ of the connection. In view of the definitions (2) and (3) of torsion and curvature, respectively, this FR implies

$$\widetilde{T}_\alpha = T_\alpha - \frac{\varepsilon}{\ell}\eta_\alpha, \qquad \widetilde{R}^{\star}_{\alpha} = R^{\star}_{\alpha} - (-1)^s \frac{\varepsilon}{2\ell} T_\alpha + (-1)^s \frac{\varepsilon^2}{4\ell^2}\eta_\alpha \quad (44)$$

for the deformed torsion and curvature, respectively. In particular, there can arise two subcases: Riemannian spacetime with deformed torsion $\widetilde{T}_\alpha = 0$, or deformed *teleparallelism* in the gauge $\widetilde{\Gamma}^{\star}_{\alpha} \stackrel{*}{=} 0$, equivalent to the covariant constraint of vanishing modified RC curvature, i.e., $\widetilde{R}^{\star}_{\alpha} = 0$.

In the latter case, coframe and connection are *Lie dual* to each other, i.e.,

$$\Gamma^{\star}_{\alpha} = (-1)^s \frac{\varepsilon}{2\ell}\vartheta_\alpha \qquad \Leftrightarrow \qquad \vartheta_\alpha = (-1)^s \frac{2\ell}{\varepsilon}\Gamma^{\star}_{\alpha}. \quad (45)$$

Observe the inversion of the parameter ε, i.e., a small deformation ε of the connection will induce a large coframe proportional to $1/\varepsilon$ and vice versa, resembling strong/weak duality. Such a duality of the *strong/weak* coupling regime of gauge fields, is the so-called *S–duality*. For Chern-Simons (super-)gravity, some of its aspects have also been discussed in Ref. [16, 20].

There could also arise the seemingly trivial case of a completely *flat* deformed spacetime, i.e., $\widetilde{T}_\alpha = 0$ and $\widetilde{R}^{\star}_{\alpha} = 0$. This would correspond to configurations with constant axial torsion and constant RC curvature as originally envision by E. Cartan, i.e.,

$$T_\alpha = \frac{\varepsilon}{\ell}\eta_\alpha, \qquad R^{\star}_{\alpha} = \frac{\rho}{\ell^2}\eta_\alpha, \quad (46)$$

where $\rho = (-1)^s\varepsilon^2/4$ depends quadratically on the deformation parameter ε.

Let us extend such ideas to supergravity in 3D: Generalizing the peculiar dynamical symmetry of BMH [2], identified as S–duality in Ref. [31], we try the following Ansatz

$$\vartheta_\alpha = (-1)^s \ell\,\Gamma^{\star}_{\alpha} + \overline{\sigma}\,\gamma_\alpha \Psi\,, \quad (47)$$

where σ is again a spinor valued zero-form and ℓ a fundamental length.

By exterior differentiation, we find

$$d\vartheta_\alpha = (-1)^s \ell d\Gamma^\star_\alpha + d(\overline{\sigma}\gamma_\alpha\Psi), \tag{48}$$

or, after separating the covariant two-forms of torsion and curvature,

$$T_\alpha - (-1)^s \eta_{\alpha\beta} \wedge \Gamma^{\star\beta} = (-1)^s \ell R^\star_\alpha - \frac{\ell}{2}\eta_{\alpha\beta\gamma}\Gamma^{\star\beta} \wedge \Gamma^{\star\gamma} + d(\overline{\sigma} \wedge \gamma_\alpha \Psi) \tag{49}$$

Let us reconstitute our Ansatz (47) in order to replace all the connection terms $\Gamma^{\star\beta}$. Then, using also the fundamental relation (18) for a Clifford algebra, we obtain

$$\begin{aligned} &T_\alpha + \frac{2}{\ell}\eta_\alpha + \frac{1}{\ell}\eta_{\alpha\beta} \wedge \overline{\sigma}\gamma^\beta\Psi \\ &= (-1)^s \ell R^\star_\alpha - \frac{1}{2\ell}\eta_{\alpha\beta\gamma}(\vartheta^\beta - \overline{\sigma}\gamma^\beta\Psi) \wedge (\vartheta^\gamma - \overline{\sigma}\gamma^\gamma\Psi) + d(\overline{\sigma}\gamma_\alpha\Psi). \end{aligned} \tag{50}$$

Now we can eliminate torsion and RC curvature via (11) and (12) with the result

$$\begin{aligned} &\frac{2}{A}\Big[(\theta_{\rm TL} + (-1)^s\theta_{\rm T})\tau^\star_\alpha - (\theta_{\rm L} + (-1)^s\theta_{\rm TL})\ell\Sigma_\alpha\Big]\ell^2 + (3 + 2\kappa - (-1)^s\rho)\eta_\alpha \\ &= -2\eta_{\alpha\beta} \wedge \overline{\sigma}\gamma^\beta\Psi - \frac{1}{2}\eta_{\alpha\beta\mu}\overline{\sigma}\gamma^\beta\Psi \wedge \overline{\sigma}\gamma^\mu\Psi + \ell d(\overline{\sigma}\gamma_\alpha\Psi). \end{aligned} \tag{51}$$

Together with (29), this leads to

$$\begin{aligned} &\frac{B}{4A} i\ell^2 \overline{\Psi}\gamma_\alpha\Psi + \frac{C}{A}\eta_\alpha \\ &= -2\eta_{\alpha\beta} \wedge \overline{\sigma}\gamma^\beta\Psi - \frac{1}{2}\eta_{\alpha\beta\mu}\overline{\sigma}\gamma^\beta\Psi \wedge \overline{\sigma}\gamma^\mu\Psi + \ell D(\overline{\sigma}\gamma_\alpha\Psi) \end{aligned} \tag{52}$$

as a condition for S-duality, where

$$B = \theta_{\rm T} + (-1)^s\theta_{\rm TL} + 2m\ell[\theta_{\rm L} + (-1)^s\theta_{\rm TL}] \tag{53}$$

and

$$C = 3A + \theta_{\rm T}[\theta_{\rm TL} + (-1)^s\theta_{\rm T}]. \tag{54}$$

In the case of vanishing B and C and in view of the massive Rarita-Schwinger equation (35), there remains a first order nonlinear differential equation for $\overline{\sigma}$ coupled to RS fields to be satisfied.

6. Membranes with torsion defects

As an example of a spacetime with torsion and/or curvature *defects* [9] or singularities, let us consider a *a planar graphene* solution within the 'mixed' MB model governed by the two Einstein-Cartan type field equations (11) and (12).

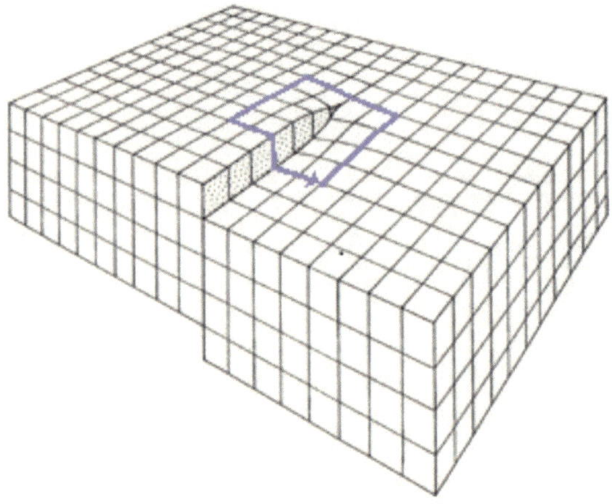

Fig. 1. 'Screw' dislocation with singular torsion in a cubic lattice. (The Cartan circuit is indicate in blue, cf. Ref. [26].]

Let us assume that the 2D membrane of a corrugated graphene is evolving in an *intrinsic* three-dimensional spacetime, suppressing for the moment the embedding of a real graphene into *flat* 4D Minkowski spacetime. Then we may adopt the convention that x^α together with y^α are spacelike orthogonal vectors which span the (x,y)–plane perpendicular to the time coordinate t, which itself is orthogonal to the world sheet of the graphene. The corresponding one–forms [29] are denoted by capital letters, i.e.

$$X := x_\alpha\, \vartheta^\alpha\,, \qquad Y := y_\alpha\, \vartheta^\alpha\,. \tag{55}$$

Moreover, the vector n^α is a timelike unit vector normal to the hypersurface with $n^\alpha\, n_\alpha = s$, the signature s of our 3D spacetime.

Following Soleng [37], cf. Anandan [1, 3, 22], we assume that the two–forms Σ_α and $\tau^\star_\alpha$ of the energy–momentum and spin current, respectively, vanish outside of the graphene sheet, whereas "inside" they are *constant*, i.e.

$$\Sigma_\alpha = \varepsilon\, x_\alpha\, X \wedge Y\,, \qquad \tau^\star_\alpha = \sigma\, y_\alpha\, X \wedge Y\,, \tag{56}$$

which satisfy

$$\vartheta^\alpha \wedge \Sigma_\alpha = 0\,, \qquad \vartheta^\alpha \wedge \tau^\star_\alpha = 0 \tag{57}$$

by construction. The constant parameters ε and σ of this *spinning string* type Ansatz are related to the exterior vacuum solution by appropriate matching conditions. For the related solution with *conical singularities* and torsion of Tod [40], we can infer that ε and σ are *delta distributions* [39] at the location of the defect, cf. Fig 1. From the specification (55) of the one–forms X and Y it can easily be inferred that the only nonzero components are $\Sigma_{\hat{0}} \neq 0$ and $\tau_{\hat{1}\hat{2}} = -\tau_{\hat{2}\hat{1}} \neq 0$.

Due to the identities (57), contractions of the second field equation (12) with x^α and y^α reveal that $x^{[\alpha}y^{\beta]}\, R_{\alpha\beta} = R_{\hat{1}\hat{2}} = -R_{\hat{2}\hat{1}} \neq 0$ are the only nonvanishing components of the RC curvature. From its covariant expression

$$R^{\alpha\beta} = \varepsilon\ell^2\, x^{[\alpha}y^{\beta]}\, X \wedge Y \tag{58}$$

there follows the identity

$$R_\beta{}^\alpha \wedge \vartheta^\beta = \frac{\varepsilon\ell^2}{2}(x^\alpha\, Y \wedge X \wedge Y - y^\alpha\, X \wedge X \wedge Y) = 0\,. \tag{59}$$

Recalling that $N^\alpha = n\rfloor\vartheta^\alpha$ is the lapse and shift vector in the (2+1)–decomposition a la ADM, the corresponding coframe and connection can now explicitly be obtained by applying a finite boost to the usual conical metric of a defect simulated by a cosmic string:

$$\begin{aligned} \vartheta^{\hat{0}} &= dt + \ell^2\sigma\rho^{*2}[1-\cos(\rho/\rho^*)]d\phi \\ \vartheta^{\hat{1}} &= d\rho\,, \qquad\qquad \vartheta^{\hat{2}} = \rho^*\sin(\rho/\rho^*)d\phi\,, \\ \Gamma^{\hat{1}\hat{2}} &= \cos(\rho/\rho^*)d\phi = -\Gamma^{\hat{2}\hat{1}}\,. \end{aligned} \tag{60}$$

From the Cartan type relation (11) and the identities (57) we can infer that the *axial* torsion

$$\mathcal{A} = {}^*(\vartheta^\alpha \wedge T_\alpha) = -(-1)^s\frac{2\kappa}{\ell^2} \tag{61}$$

of such a membrane defect is a non-vanishing constant. Thus, in 3D there is no contribution to the Pointrjagin type term $d(\mathcal{A} \wedge d\mathcal{A})$ from the axial torsion.

Moreover, the Nieh–Yan term dC_{T} proportional to $d\,{}^*\mathcal{A}$ *vanishes identically* for this example of a spinning cosmic string exhibiting a *torsion line defect*.

7. Outlook: Graphene and supersymmetry

Fundamental interactions are rather successful formulated in terms of Yang-Mills theories with large gauge groups, stipulating that symmetry breaking is occurring in the ground state. The idea of supersymmetry or supergravity, anticipated to some extent already by Hermann Weyl [42], goes in the same direction but so far lacks empirical support in particle physics.

Recently, graphene [33] as a new material has attracted a lot of attention because its charge carriers can be described by massless Dirac fields, cf. Ref. [41], whereas the flexural models of the 2D membrane of graphene have been tentatively considered as membranes, cf. Ref. [25], evolving in $2+1$ dimensional curved, but conformally flat spacetime [24]. There are also indications of dislocations [9] related to torsion.

A related topological framework with a coupling to Dirac fields in 3D has been considered before by Lemke and Mielke [27]. It seems to be feasible to enlarge the dynamical framework of the theory by including supersymmetry, cf. Ref. [17] and apply the topological ideas developed to some extent in this paper.

8. Acknowledgments

We would like to thank to Friedrich W. Hehl for constructive comments. Moreover, (E.W.M.) acknowledges the support of the SNI and thanks Noelia, Miryam Sophie Naomi and Markus Gérard Erik, for encouragement.

9. Appendices

A: Variations of Chern–Simons terms

Gauging the Poincaré group in (2+1) dimensions, local translations and (Lorentz-) rotations give rise to two type of gauge potentials, the coframe ϑ^γ and the dual of Lorentz-connection $\Gamma^\star_\alpha$. Then the two Bianchi identities of Riemann-Cartan geometry can be rewritten as

$$DT^\alpha \equiv (-1)^s\, \eta^{\alpha\beta} \wedge R^\star_\beta\,, \tag{62}$$

$$DR^\star_\alpha \equiv 0. \tag{63}$$

In 3D the corresponding Chern–Simons three–forms of gauge type $C = \mathrm{Tr}\{A \wedge F\}$, are

$$C_{\mathrm{T}} := \frac{1}{2\ell^2}\vartheta^\alpha \wedge T_\alpha = -\frac{(-1)^s}{\ell^2}\eta^\alpha \wedge K^\star_\alpha, \quad C_{\mathrm{L}} := (-1)^s\,\Gamma^{\star\alpha} \wedge R^\star_\alpha - \frac{1}{3!}\,\eta_{\alpha\beta\gamma}\,\Gamma^{\star\alpha} \wedge \Gamma^{\star\beta} \wedge \Gamma^{\star\gamma}. \tag{64}$$

and

$$C_{\mathrm{TL}} := \frac{1}{\ell}\left(\Gamma^{\star\alpha} \wedge T_\alpha - \frac{(-1)^s}{2}\,\eta_{\alpha\beta\gamma}\,\Gamma^{\star\alpha} \wedge \Gamma^{\star\beta} \wedge \vartheta^\gamma\right). \tag{65}$$

The variational derivatives of these terms lead us to the following expressions

$$\frac{\delta C_{\mathrm{T}}}{\delta\vartheta^\alpha} = \frac{1}{\ell^2}\,T_\alpha \qquad \frac{\delta C_{\mathrm{T}}}{\delta\Gamma^{\star\alpha}} = \frac{(-1)^s}{\ell^2}\,\eta_\alpha\,, \tag{66}$$

$$\frac{\delta C_{\mathrm{L}}}{\delta\vartheta^\alpha} = 0, \qquad \frac{\delta C_{\mathrm{L}}}{\delta\Gamma^{\star\alpha}} = (-1)^s\, 2R^\star_\alpha, \tag{67}$$

$$\frac{\delta C_{\mathrm{TL}}}{\delta\vartheta^\alpha} = \frac{1}{\ell}R^\star_\alpha\,, \qquad \frac{\delta C_{\mathrm{TL}}}{\delta\Gamma^{\star\alpha}} = \frac{1}{\ell}T_\alpha, \tag{68}$$

respectively. Note that these three–forms are uniquely related to the torsion T_α, the curvature $R^\star_\alpha$, and the cosmological term η_α, as developed in much more detail in Ref. [21].

B: The η–basis for exterior forms in 3D

The symbol $\wedge$ denotes the exterior product of forms, the symbol $\rfloor$ the interior product of a vector with a form and * the Hodge star (or left dual) operator which maps a p–form into a $(3-p)$–form. It has the property that

$$^*{}^*\Phi^{(p)} = (-1)^{p(3-p)+s}\Phi^{(p)}, \tag{69}$$

where p is the degree of the form Φ and s denotes the number of negative *eigenvalues* of the metric, i.e., the signature of spacetime.

The volume three–form is defined by

$$\eta := \frac{1}{3!}\,\eta_{\alpha\beta\gamma}\,\vartheta^\alpha \wedge \vartheta^\beta \wedge \vartheta^\gamma, \tag{70}$$

where $\eta_{\alpha\beta\gamma} := \sqrt{|\det g_{\mu\nu}|}\,\epsilon_{\alpha\beta\gamma}$, and $\epsilon_{\alpha\beta\gamma}$ is the Levi–Civita symbol. The forms $\{\eta, \eta_\alpha, \eta_{\alpha\beta}, \eta_{\alpha\beta\gamma}\}$ span a *dual basis* for the algebra of arbitrary p–forms in 3D, where

$$\begin{aligned}
\eta_\alpha &:= e_\alpha \rfloor \eta = \frac{1}{2}\eta_{\alpha\beta\gamma}\vartheta^\beta \wedge \vartheta^\gamma = {}^*\vartheta_\alpha, \\
\eta_{\alpha\beta} &:= e_\beta \rfloor \eta_\alpha = \eta_{\alpha\beta\gamma}\vartheta^\gamma = {}^*(\vartheta_\alpha \wedge \vartheta_\beta), \\
\eta_{\alpha\beta\gamma} &:= e_\gamma \rfloor \eta_{\alpha\beta}.
\end{aligned} \tag{71}$$

In 3D, the following relations for the η–basis hold:

$$\begin{aligned}
\eta^{\alpha\beta\gamma}\eta_{\alpha\beta\gamma} &= (-1)^s 3!, \\
\eta^{\alpha\beta\gamma}\eta_{\alpha\beta\nu} &= (-1)^s 2\delta^\gamma_\nu, \\
\eta^{\alpha\beta\gamma}\eta_{\alpha\mu\nu} &= (-1)^s \delta^\beta_\mu \delta^\gamma_\nu = (-1)^s 2\delta^\beta_{[\mu}\delta^\gamma_{\nu]}, \\
\eta^{\alpha\beta\gamma}\eta_{\rho\mu\nu} &= (-1)^s \delta^{\alpha\beta\gamma}_{\rho\mu\nu},
\end{aligned} \tag{72}$$

and

$$\eta_\beta \wedge \eta^{\alpha\beta} = e_\beta \rfloor (\eta \wedge \eta^{\alpha\beta}) + \eta \wedge e_\beta \rfloor \eta^{\alpha\beta} \equiv 0 \tag{73}$$

due to the antisymmetry of $\eta^{\alpha\beta}$ and the fact that $\eta \wedge \eta^{\alpha\beta}$ would already be a four-form in 3D.

C: Identities for spinor–valued forms

Now some relations of special importance are presented which take care of the order of the forms in the exterior products and its Dirac adjoint: We would like to remind the reader that Φ is a p–form and Ψ a q–form with the spinor indices suppressed:

$$\Phi^p \wedge \Psi^q = (-1)^{p\cdot q}\Psi^q \wedge \Phi^p, \tag{74}$$

$$\overline{\Phi^p \wedge \Psi^q} = (-1)^{p\cdot q}\,\overline{\Psi^q} \wedge \overline{\Phi^p}, \tag{75}$$

$$\Phi^p \wedge {}^*\Psi^p = \Psi^p \wedge {}^*\Phi^p, \tag{76}$$

$$e_\alpha \rfloor (\Phi^p + \Psi^q) = e_\alpha \rfloor \Phi^p + e_\alpha \rfloor \Psi^q, \tag{77}$$

$$e_\alpha \rfloor (\Phi^p \wedge \Psi^q) = (e_\alpha \rfloor \Phi^p) \wedge \Psi^q + (-1)^p \Phi^p \wedge (e_\alpha \rfloor \Psi^q), \tag{78}$$

$$\vartheta^\alpha \wedge (e_\alpha \rfloor \Phi) = p\Phi \tag{79}$$

$${}^*(\Phi \wedge \vartheta_\alpha) = e_\alpha \rfloor {}^*\Phi, \tag{80}$$

$$\overline{\gamma} = \gamma. \tag{81}$$

D: No axial torsion restrictions in 3D

Spaces of constant curvature deserve special attention in General Relativity, in particular in the cosmological context. In particular, when the RC curvature is constant as in Eq. (10), i.e.

$$R^\star_\alpha = \frac{\rho}{\ell^2}\eta_\alpha = \frac{\rho}{2\ell^2}\eta_{\alpha\beta\gamma}\vartheta^\beta \wedge \vartheta^\gamma, \tag{82}$$

the Bianchi identities (62) and (63) could lead to constraints on the admissible torsion T^α, as in 4D and higher dimensions. However, in 3D the situation is different: Using Appendix B, the first Bianchi identity yields

$$\begin{aligned}(-1)^s \eta^{\alpha\beta} \wedge R^*_\beta &= (-1)^s \frac{\rho}{\ell^2} \eta_\beta \wedge \eta^{\alpha\beta} = (-1)^s \frac{\rho}{2\ell^2} \left(\eta_{\beta\mu\nu}\eta^{\alpha\beta\gamma}\right) \vartheta_\gamma \wedge \vartheta^\mu \wedge \vartheta^\nu \\ &= -(-1)^s 2\delta^\alpha_{[\mu}\delta^\gamma_{\nu]}\, \vartheta_\gamma \wedge \vartheta^\mu \wedge \vartheta^\nu = 0.\end{aligned} \tag{83}$$

Furthermore, the exterior covariant derivative of Eq. (10) provides the identity

$$DT_\alpha = \frac{2\kappa}{\ell} D\eta_\alpha = \frac{2\kappa}{\ell} T^\beta \wedge \eta_{\alpha\beta} = \frac{4\kappa^2}{\ell^2} \eta_{\alpha\beta} \wedge \eta^\beta \equiv 0. \tag{84}$$

Thus the first Bianchi identity does not give any further information. The second Bianchi identity (63) yields

$$DR^\star_\alpha = \frac{\rho}{\ell^2} D\eta_\alpha = \frac{2\kappa\rho}{\ell^3} \eta_{\alpha\beta} \wedge \eta^\beta \equiv 0 \tag{85}$$

which is identically zero by a similar argument, or by employing Eq. (73). Consequently, the Bianchi identities impose *no* restrictions on the axial torsion given by (10) in 3D, a fact which has allowed us to construct something non-trivial from the MB model.

10. References

[1] Anandan, J. (1994). Topological and geometrical phases due to gravitational field with curvature and torsion, *Physics Letters* A195, 284–292; (1996). Gravitational phase operator and cosmic strings, *Physical Review* D53, 779–786.

[2] Baekler, P., Mielke, E. W. & Hehl, F. W. (1992). Dynamical symmetries in topological 3D gravity with torsion, *Il Nuovo Cimento* B107, 91–110.

[3] Bakke, K., Furtado, C. & Nascimento, J. R. (2009). Gravitational geometric phase in the presence of torsion, *The European Physical Journal* C60, 501–507.

[4] Cacciatori, S. L., Caldarelli, M. M., Giacomini, A., Klemm, D. & Mansi, D. S. (2006). Chern-Simons formulation of three-dimensional gravity with torsion and nonmetricity, *Journal Geometrical Physics* 56, 2523–2543.

[5] Carlip, S. (1995). Lectures on (2+1)–dimensional gravity, *Journal of Korean Physical Society* 28, S447–S467.

[6] Carlip, S. (1998). *Quantum gravity in* $(2+1)$*–dimensions*, Cambridge University Press.

[7] Carlip, S. (1995). The $(2+1)$–dimensional black hole, *Classical Quantum Gravitation* 12, 2853–2880; (2005). Conformal field theory, (2+1)-dimensional gravity, and the BTZ black hole, *Classical Quantum Gravitation* 22, R85–R124.

[8] Chern, S. S. & Simons, J. (1971). Some cohomology classes in principal fiber bundles and their application to Riemannian geometry, *Proceedings of Natural Academy of Science* 68, 791–794.

[9] de Juan, F. , Cortijo, A. & Vozmediano, M. A. H. (2010). Dislocations and torsion in graphene and related systems, *Nuclear Physics* B828, 625–637 .

[10] Dereli, T. & Tucker, R. W. (1988). Gravitational interactions in 2+1 dimensions, *Classical Quantum Gravitation* 5, 951–959.

[11] Dereli, T. & Verçin, A. (1991). A gauge model of amorphous solids containing defects II. Chern-Simons free energy, *Philosophy Magnetics* B64, 509.
[12] Deser, S., Jackiw, R. & Templeton, S. (1982). Topologically massive gauge theories, *Annals of Physics* 140, 372–411.
[13] Deser, S. (1984). Cosmological topological supergravity, *Quantum Theory of Gravity*, essays in honor of the 60th Birthday of Bryce S. DeWitt. S. M. Chistensen, ed. Adam Hilger, Bristol, 374.
[14] Deser, S. (1989). Three topics in three dimensions, *Supermembranes and Physics in 2+1 Dimensions*. Proc of the Trieste Conference. ICTP, Trieste, M. J. Duff, C. N. Pope and E. Sezgin, eds. World Scientific, Singapore, p.239.
[15] Deser, S. & Zumino, B. (1976). Consistent supergravity, *Physics Letters* B62, 335.
[16] Deser, S. & McCarthy, J. G. (1990). Selfdual formulations of $D = 3$ gravity theories, *Physics Letters* B246, 441. [Addendum-ibid. B248, 473].
[17] Ezawa, M. (2008). Supersymmetric structure of quantum Hall effects in graphene, *Physics Letters* A 372, 924.
[18] Falbo–Kenkel, M. K. & Mansouri, F. (1993). Nonperturbative two–body dynamics in 2+1–dimensional gravity, *Journal of Mathematical Physics*. 34, 139–153.
[19] Freedman, D. Z. (1993). Some beautiful equations of mathematical physics, *Dirac Lecture Trieste*, preprint CERN, TH 7367/94.
[20] Garcia-Compean, H., Obregon, O., Ramirez, C. & Sabido, M. (2001). On S-duality in (2+1)-Chern-Simons supergravity, *Physical Review* D64, 024002.
[21] Hehl, F. W., McCrea, J. D., Kopczynski, W. & Mielke, E. W. (1991). Chern–Simons terms in metric–affine spacetime: Bianchi identities as Euler–Lagrange equations, *Journal of Mathematical Physics* 32, 2169–2180.
[22] Hehl, F. W., McCrea, J. D., Mielke, E. W. & Ne'eman, Y. (1995). Metric–affine gauge theory of gravity: Field equations, Noether identities, world spinors, and breaking of dilation invariance, *Physics Report* 258, 1–171.
[23] Howe, P. S. & Tucker, R. W. (1978). Local supersymmetry in (2+1)-dimensions. 1. Supergravity and differential forms, *Journal of Mathematical Physics* 19, 869–873.
[24] Iorio, A. (2011). Weyl-gauge symmetry of graphene, *Annals of Physics* 326, 1334–1353.
[25] Kerner, R. & Naumis, G. (2011). Interaction of flexural phonons with electrons in graphene: A generalized Dirac equation in corrugated surfaces, [arXiv:1102.0746 [cond-mat.mtrl-sci]].
[26] Lazar, M. & Hehl, F. W. (2010). Cartan's spiral staircase in physics and, in particular, in the gauge theory of dislocations, *Foundations of Physics* 40, 1298–1325.
[27] Lemke, J. & Mielke, E. W. (1993). Gravitational moments of spin one–half particles and of topological massive photons and gravitons in $2 + 1$ dimensions, *Physics Letters* A175, 277–281.
[28] Mielke, E. M. & Baekler, P. (1991). Topological gauge model of gravity with torsion, *Physics Letters* A156, 399.
[29] Mielke, E. M. & Kreimer, D. (1998). Chiral anomaly in Ashtekar's approach to canonical gravity, *International Journal of Modern Physics* D7, 535–548.
[30] Mielke, E. W. & Macías, A. (1999). Chiral supergravity and anomalies, *Annalen Physik* 8, 301.
[31] Mielke, E. W. & Maggiolo, A. R. (2003). Rotating black hole solution in a generalized topological 3D gravity with torsion, *Physical Review* D68, 104026.

[32] Mielke, E. W. & Maggiolo, A. R. (2007). S-duality in 3D gravity with torsion, *Annals of Physics* (N.Y.) 322, 341–362.
[33] Novoselov, K. S., Geim, A. K., Morozov, S. V., Jiang, D., Katsnelson, M. I., Grigorieva, I. V., Dubonos, S. V. & Firsov, A. A. (2005). Two-dimensional gas of massless Dirac fermions in graphene, *Nature* 438, 197.
[34] Nieuwenhuizen, P. van (1981). Supergravity, *Physics Reports* 68, 189.
[35] Rarita, W. & Schwinger, J. S. (1941). On a theory of particles with half integral spin, *Physical Review* 60, 61.
[36] Schroers, B. J. (2011). Quantum gravity and non-commutative spacetimes in three dimensions: a unified approach, arXiv:1105.3945v1 [gr-qc] 1–22.
[37] Soleng, H. H. (1992). A spinning string, *General Relativity and Gravitation* 24, 111–117.
[38] Staruszkiewicz, A. (1963). Gravitation theory in three-dimensional space, *Acta Physica Polonica* 24, 735–740.
[39] Taub, A. H. (1980). Space-times with distribution valued curvature tensors, *Journal of Mathematical Physics* 21, 1423–1431.
[40] Tod, K. P. (1994). Conical singularities and torsion, *Classical Quantum and Gravitation* 11, 1331.
[41] Villalba, V. M. & Maggiolo, A. R. (2001). Energy spectrum of a 2-D Dirac oscillator in the presence of a constant magnetic field, *The European Physical Journal* B22, 31–35.
[42] Weyl, H. (1931). Geometrie und Physik, *Naturwissenschaften* 19, 49.
[43] Witten, E. (1988). (2+1)-dimensional gravity as an exactly soluble system, *Nuclear Physics* B311, 46.

4

Quantum Gravity Insights from Smooth 4-Geometries on Trivial $\mathbb{R}^4$

Jerzy Król
Institute of Physics, University of Silesia, Katowice
Poland

1. Introduction

The search for the theory of quantum gravity (QG) in 4-dimensions (4D) is one of the most significant challenges of temporary physics. The great effort and insights of many theoreticians and experimentalists resulted in the emergence of one of the greatest achievements of 20th century science, i.e. standard model of particles and fields (SM). SM (with its minimal extensions by massive neutrinos and after renormalization) describes and predicts, with enormous accuracy, almost all perturbative aspects and behaviour of interacting quantum fields and particles which place themselves in the realm of electromagnetic, strong and weak nuclear interactions, within the range of energies up to few TeV. However, gravity at quantum level is not covered by this pattern. The oldest, semiclassical, approach to QG relies on the quantization of metric field which is understood as the perturbation of the ground spacetime metric. This is exactly in the spirit of quantum field theory (QFT) as in SM. There should follow various correlation functions of physical processes where gravity at quantum level is present. There should, but actually they do not since the expressions are divergent and the theory is not renormalizable. Even the presence of supersymmetry does not change this substantially. On the other hand, we have a wonderful theory of general relativity (GR) which, however, is a theory of classical gravity and it prevents its quantization in 4D.

Among existing approaches to QG, superstring theory is probably the most advanced and conservative one. It attempts to follow GR and quantum mechanics as much as possible. However, superstring theory has to be formulated in 10 spacetime dimensions and on fixed, not dynamical, background. Many proposals how to reach the observed physics from 10D superstrings were worked out within the years. These are among others, compactiffication, flux stabilization, brane configuration model-buildings, brane worlds, holography or anti-de-Sitter/conformal field theory duality, i.e. AdS/CFT. There exists much ambiguity, however, with determining 4D results by these methods. Some authors estimate that there exist something about 10^{500} different backgrounds of superstring theory which all could be "good" candidates expressing 4D physics. This means that similar variety of possible models for true physics is predicted by superstring theory. To manage with such huge amount of "good" solutions, there was proposed to use the methods of statistical analysis to such *landscape* of possible backgrounds. Anyway, one could expect better prediction power from the fundamental theory which would unify gravity with other interactions at quantum level. On the other hand, superstring theory presents beautiful, fascinating and extremly rich mathematics which is still not fully comprehended.

Therefore Asselmeyer-Maluga & Król (2010) have proposed recently how to find connections of superstring theory with dimension 4 in a new way not relying on the standard techniques. Whole the approach derives from mathematics, especially low dimensional differential topology and geometry. In that approach one considers superstring theory in 10D and supersymmetry as "merely" mathematics describing directly, at least in a variety of important cases, the special smooth geometry on Euclidean, topologically trivial, manifold $\mathbb{R}^4$. These are various non-diffeomorphic different smooth structures. Smooth manifold $\mathbb{R}^4$ with such non-standard smoothness is called exotic $\mathbb{R}^4$ and as a smooth Riemannian manifold allows for a variety of metrics. This exotic geometry in turn, is regarded as underlying smoothness for 4-spacetime and is directly related to physics in dimension 4.

The way towards crystallizing such point of view on string theory was laborious and required many important steps. The breakthrough findings in differential topology, from the eighties of the previous century, showed that indeed there are different from the standard one, smoothings of the simplest Euclidean 4-space (see e.g. Asselmeyer-Maluga & Brans (2007)). Spacetime models usually are based on 4D smooth manifolds, hence they are locally described with respect to the standard smooth $\mathbb{R}^4$. Anything what happens to this fundamental building block might be important at least to classical physics formulated on such spacetime. Indeed, it was conjectured by Brans (1994a;b), and then proved by Asselmeyer (1996) and Sładkowski (2001), that exotic smooth $\mathbb{R}^4$'s can act as sources for the external gravitational field in spacetime. Even mathematics alone, strongly distinguishes these smooth open 4-manifolds: among all $\mathbb{R}^n$ only the case $n = 4$ allows for different smoothings of Euclidean $\mathbb{R}^n$. For any other $\mathbb{R}^n$, $n \neq 4$ there exists unique smooth structure. Moreover, there exists infinitely continuum many different smoothings for $\mathbb{R}^4$. However, mathematical tools suitable for the direct description of, say, metrics or functions on exotic $\mathbb{R}^4$ are mostly unknown (see however Asselmeyer-Maluga & Brans (2011)). The main obstruction which prevents progress in our understanding of exotic smoothness on $\mathbb{R}^4$ is that there is no known effective coordinate presentation. As the result, no exotic smooth function on any such $\mathbb{R}^4$ is known, even though there exist infinite continuum many different exotic $\mathbb{R}^4$. Such functions are smooth in the exotic smoothness structure, but fail to be differentiable in a standard way determined by the topological product of axes. This is also a strong limitation for the applicability of the structures to physics. Let us note that smooth structures on open 4-manifolds, like on $\mathbb{R}^4$, are of special character and require special mathematics which, in general, is not completely understood now. The case of *compact* 4-manifolds and their smooth structures is much better recognized also from the point of view of physics (see e.g. Asselmeyer-Maluga (2010); Asselmeyer-Maluga & Brans (2007); Witten (1985)). The famous exception is, however, not resolved yet, negation of the 4D Poincaré conjecture stating that there exists exotic S^4.

Bižaca (1994) constructed an infinite coordinate patch presentation by using Casson handles. Still, it seems hopeless to extract physical information from that. The proposition by Król (2004a;b; 2005) indicated that one should use methods of set theory, model theory and categories to grasp properly some results relevant to quantum physics. Such low level constructions modify the smoothness on $\mathbb{R}^4$ and the structures survive the modifications as a *classical* exotic $\mathbb{R}^4$. Thus, functions, although from different logic and category, approach exotic smooth ones, such that some quantum structures emerge due to the rich categorical formalism involved. Still, to apply exotic 4-smoothness in variety of situations one needs more direct relation to existing calculus.

Even though neither any explicit exotic metric nor the function on $\mathbb{R}^4$ is known, recent relative results made it possible to apply these exotic structures in a variety of contexts relevant to physics. In particular, strong connection with quantum theories and quantization was shown by Asselmeyer-Maluga & Król (2011b). First, we deal here exclusively with *small* exotic $\mathbb{R}^4$. These arise as the result of failing h-cobordism theorem in 4D (see e.g. Asselmeyer-Maluga & Brans (2007)). The others, so called large exotic $\mathbb{R}^4$, emerge from failing the smooth surgery in 4D. Second, the main technical ingredient of the relative approach to small 4-exotics is the relation of these with some structures defined on a 3-sphere. This S^3 should be placed as a part of the boundary of some contractible 4-submanifold of $\mathbb{R}^4$. This manifold is the Akbulut cork and its boundary is, in general, a closed 3-manifold which has the same homologies as ordinary 3-sphere – homology 3-sphere. Next, we deal with the parameterized by the radii $\rho \in \mathbb{R}$ of S^4 as a subset of R^4, a family of exotic $\mathbb{R}^4_\rho$ each of which is the open submanifold of standard $\mathbb{R}^4$. This is the *radial family* of small exotic $\mathbb{R}^4$'s or the deMichellis-Freedman family DeMichelis & Freedman (1992). Let CS be the standard Cantor set as a subset of $\mathbb{R}$, then the crucial result is:

Theorem 1 (Asselmeyer-Maluga & Król (2011b)). *Let us consider a radial family R_t of small exotic $\mathbb{R}^4_t$ with radius ρ and $t = 1 - \frac{1}{\rho} \subset CS \subset [0,1]$ induced from the non-product h-cobordism W between M and M_0 with Akbulut cork $A \subset M$ and $A \subset M_0$, respectively. Then, the radial family R_t determines a family of codimension-one foliations of ∂A with Godbillon-Vey invariant ρ^2. Furthermore, given two exotic spaces R_t and R_s, homeomorphic but non-diffeomorphic to each other (and so $t \neq s$), then the two corresponding codimension-one foliations of ∂A are non-cobordant to each other.*

This theorem gives a direct relation of small exotic $\mathbb{R}^4$'s - from the radial family, and (codimension one) foliations of some S^3 - from the boundary of the Akbulut cork. M and M_0 are compact non-cobordant 4-manifolds, resulting from the failure of the 4D h-cobordism theorem (see the next section). Such relativization of 4-exotics to the foliations of S^3 is the source of variety of further mathematical results and their applications in physics. One example of these is the quantization of electric charge in 4D where instead of magnetic monopoles one considers exotic smoothness in some region in spacetime Asselmeyer-Maluga & Król (2009a). More examples of this kind will be presented in the course of this Chapter.

Now we are ready to formulate two important questions as guidelines for this work:

i. What if smooth structure, with respect to which standard model of particles is defined, is not the "correct" one and it does not match with the smooth structure underlying GR and theories of quantum gravity, in 4 dimensions?

ii. What if particles and fields, as in standard model of particles, are not fundamental from the point of view of gravity in 4 dimensions? Rather, more natural are effective condensed matter states, and these states should be used in order to unify quantum matter with general relativity.

This Chapter is thought as giving the explanation for the above questions and for the existence of a fundamental connection between these, differently looking, problems. New point of view on the reconciliation of quantum field theory with general relativity in 4 physical dimensions, emerges. The exact description of quantum matter and fields coupled with QG in 4D, at least in some important cases, is presented. The task to build a final theory of QG in 4D is thus seen from different perspective where rather effective states of condensed matter are well suited for the reconciliation with QG. Such approach is also motivated by the AdS/CFT dualities where effective matter states (without gravity) are described by dual theories with gravity. Hence gravity is inherently present in description of such condensed matter states.

In the next section we describe the relation of small exotic $\mathbb{R}^4$ with foliations of S^3 and WZW models on $SU(2)$. Then we show the connections between string theory and exotic $\mathbb{R}^4$. In particular 4-smoothness underlying spacetime emerges from superstring calculations and it modifies the spectra of charged particles in such spacetime. In Sec. 4 we discuss the Kondo state and show that it generates the same exotic 4-smoothness. Moreover, the Kondo state, when survive the high energy and relativistic limit, would couple to the gravity backgrounds of superstring theory. The backgrounds are precisely those related with exotic smooth $\mathbb{R}^4$ as in Sec. 3. We conjecture that one could encounter the experimental trace of existence of exotic $\mathbb{R}^4_p$ in the k-channel, $k > 2$, Kondo effect, where the usual fusion rules of the $SU(2)_k$ WZW model would be modified to these of $SU(2)_p$ WZW in high energies.

Next in Sec. 5 we present the connections of branes configurations in superstring theory with non-standard 4-smoothness of $\mathbb{R}^4$. Discussion and conclusions close the Chapter.

2. Foliations, WZW σ-models and exotic $\mathbb{R}^4$

An exotic $\mathbb{R}^4$ is a topological space with $\mathbb{R}^4$−topology but with a smooth structure different (i.e. non-diffeomorphic) from the standard $\mathbb{R}^4_{std}$ obtaining its differential structure from the product $\mathbb{R} \times \mathbb{R} \times \mathbb{R} \times \mathbb{R}$. The exotic $\mathbb{R}^4$ is the only Euclidean space $\mathbb{R}^n$ with an exotic smoothness structure. The exotic $\mathbb{R}^4$ can be constructed in two ways: by the failure to split arbitrarily a smooth 4-manifold into pieces (large exotic $\mathbb{R}^4$) and by the failure of the so-called smooth h-cobordism theorem (small exotic $\mathbb{R}^4$). Here, we deal with the later kind of exotics. We refer the reader to Asselmeyer-Maluga & Brans (2007) for general presentation of various topological and geometrical constructions and their physical perspective. Another useful mathematical books are Gompf & Stipsicz (1999); Scorpan (2005). The reader can find further results in original scientific papers.

Even though there are known, and by now rather widely discussed (see the Introduction), difficulties with making use of different differential structures on $\mathbb{R}^4$ (and on other open 4-manifolds) in explicit coordinate-like way (see e.g. Asselmeyer-Maluga & Brans (2007)), it was, however, established, in a series of recent papers, the way how to relate these 4-exotics with some structures on S^3 (see e.g. Asselmeyer-Maluga & Król (2009a;b; 2011b)). This S^3 is supposed to fulfil specific topological conditions: it has to lie in ambient $\mathbb{R}^4$ such that it is a part of the boundary of some compact 4-submanifold with boundary, i.e. Akbulut cork. If so, one can prove that exotic smoothness of the $\mathbb{R}^4$ is tightly related to codimension-one foliations of this S^3, hence, with the 3-rd real cohomology classes of S^3. Reformulating Theorem 1 we have [Asselmeyer-Maluga & Król (2009a)]:

The exotic $\mathbb{R}^4$'s, from the radial family of exotic $\mathbb{R}^4$'s embedded in standard $\mathbb{R}^4$, are determined by the codimension-1 foliations, $\mathcal{F}$'s, with non-vanishing Godbillon-Vey (GV) class in $H^3(S^3, \mathbb{R})$ of a 3-sphere lying at the boundary of the Akbulut corks of $\mathbb{R}^4$'s. The radius in the family, ρ, and value of GV are related by $\mathrm{GV} = \rho^2$. We maintain: the exoticness is localized at a 3-sphere inside the small exotic $\mathbb{R}^4$ (seen as a submanifold of $\mathbb{R}^4$).

Let us explain briefly, following Asselmeyer-Maluga & Król (2011d), how the codimension-1 foliations of S^3 emerges from the structure of exotic $\mathbb{R}^4$. The complete construction and proof can be found in Asselmeyer-Maluga & Król (2009a).

Small exotic $\mathbb{R}^4$ is determined by the compact 4-manifold A with boundary ∂A which is homology 3-sphere, and attached several Casson handles CH's. A is the Akbulut cork and CH is built from many stages towers of immersed 2-disks. These 2-disks cannot be embedded and the intersection points can be placed in general position in 4D in separated double points. Every CH has infinite many stages of intersecting disks. However, as Freedman proved, CH is topologically the same as (homeomorphic to) open 2-handle, i.e. $D^2 \times \mathbb{R}^2$. Now if one replaces CH's, from the above description of small exotic $\mathbb{R}^4$, by ordinary open 2-handles (with suitable linking numbers in the attaching regions) the resulting object is standard $\mathbb{R}^4$. The reason is the existence of infinite (continuum) many diffeomorphism classes of CH, even though all are topologically the same.

Consider the following situation: one has two topologically equivalent (i.e. homeomorphic), simple-connected, smooth 4-manifolds M, M', which are not diffeomorphic. There are two ways to compare them. First, one calculates differential-topological invariants like Donaldson polynomials Donaldson & Kronheimer (1990) or Seiberg-Witten invariants Akbulut (1996). But there is yet another possibility – one can change a manifold M to M' by using a series of operations called surgeries. This procedure can be visualized by a 5-manifold W, the cobordism. The cobordism W is a 5-manifold having the boundary $\partial W = M \sqcup M'$. If the embedding of both manifolds M, M' into W induces a homotopy-equivalence then W is called an h-cobordism. Moreover, we assume that both manifolds M, M' are compact, closed (without boundary) and simply-connected. Freedman (1982) showed that every h-cobordism implies a homeomorphism, hence h-cobordisms and homeomorphisms are equivalent in that case. Furthermore, the following structure theorem for such h-cobordisms holds true [Curtis & Stong (1997)]:
Let W be a h-cobordism between M, M'. Then there are contractable submanifolds $A \subset M, A' \subset M'$ together with a sub-cobordism $V \subset W$ with $\partial V = A \sqcup A'$ (the disjoint oriented sum), so that the h-cobordism $W \setminus V$ induces a diffeomorphism between $M \setminus A$ and $M' \setminus A'$.
Thus, the smoothness of M is completely determined (see also Akbulut & Yasui (2008; 2009)) by the contractible submanifold A (Akbulut cork) and its embedding $A \hookrightarrow M$ determined by a map $\tau : \partial A \to \partial A$ with $\tau \circ \tau = id_{\partial A}$ and $\tau \neq \pm id_{\partial A}$($\tau$ is an involution). Again, according to Freedman (1982), the boundary of every contractible 4-manifold is a homology 3-sphere. This h-cobordism theorem is employed to construct an exotic $\mathbb{R}^4$. First, one considers a neighborhood (tubular) of the sub-cobordism V between A and A'. The interior of V, $int(V)$, (as open manifold) is homeomorphic to $\mathbb{R}^4$. However, if (and only if) M and M' are not diffeomorphic (still being homeomorphic), then $int(V) \cap M$ is an exotic $\mathbb{R}^4$.

Next, Bižaca (1994) and Bižaca & Gompf (1996) showed how to construct an explicit handle decomposition of the exotic $\mathbb{R}^4$ by using $int(V)$. The details of the construction can be found in their papers or in the book Gompf & Stipsicz (1999). The idea is simply to use the cork A and add some Casson handle to it. The interior of this resulting structure is an exotic $\mathbb{R}^4$. The key feature here is the appearance of CH. Briefly, a Casson handle CH is the result of attempts to embed a disk D^2 into a 4-manifold. In most cases this attempt fails and Casson (1986) searched for a possible substitute, which is just what we now call a Casson handle. Freedman (1982) showed that every Casson handle CH is homeomorphic to the open 2-handle $D^2 \times \mathbb{R}^2$ but in nearly all cases it is not diffeomorphic to the standard handle, Gompf (1984; 1989). The Casson handle is built by iteration, starting from an immersed disk in some 4-manifold M, i.e. a map $D^2 \to M$ which has injective differential. Every immersion $D^2 \to M$ is an embedding except on a countable set of points, the double points. One can "kill" one double point by immersing another disk into that point. These disks form the first stage of the Casson handle. By iteration one can produce the other stages. Finally, we consider a tubular neighborhood $D^2 \times D^2$ of this immersed disk, called a kinky handle, on each stage. The union of all neighborhoods of all stages is the Casson handle. So, there are two input data involved with the construction of a CH: the number of double points in each stage and their orientation $\pm$. Thus, we can visualize the Casson handle CH by a tree: the root is the immersion $D^2 \to M$ with k double points, the first stage forms the next level of the tree with k vertices connected with the root by edges etc. The edges are evaluated using the orientation $\pm$. Every Casson handle can be represented by such an infinite tree. The structure of CH as immersed many-layers 2-disks will be important in Sec. 4 where we will assign fermion fields to CH's.

Next, we turn again to the radial family of small exotic $\mathbb{R}^4$, i.e. a continuous family of exotic $\{\mathbb{R}^4_\rho\}_{\rho\in[0,+\infty]}$ with parameter ρ so that $\mathbb{R}^4_\rho$ and $\mathbb{R}^4_{\rho'}$ are non-diffeomorphic for $\rho \neq \rho'$. The point is that this radial family has a natural foliation (see Theorem 3.2 in DeMichelis & Freedman (1992)) which can be induced by a polygon P in the two-dimensional hyperbolic space $\mathbb{H}^2$. The area of P is a well-known invariant, the Godbillon-Vey class as the element in $H^3(S^3,\mathbb{R})$. Every GV class determines a codimension-one foliation on the 3-sphere (firstly constructed by Thurston (1972); see also the book Tamura (1992) chapter VIII for the details). This 3-sphere is a part of the boundary ∂A of the Akbulut cork A (there is an embedding $S^3 \to \partial A$). Furthermore, one can show that the codimension-one foliation of the 3-sphere induces a codimension-one foliation of ∂A so that the area of the corresponding polygons (and therefore the foliation invariants) agree. The Godbillon-Vey invariant $[GV] \in H^3(S^3,\mathbb{R})$ of the foliation is related to the parameter of the radial family by $\langle GV, [S^3]\rangle = \rho^2$ using the pairing between cohomology and homology (the fundamental homology class $[S^3] \in H_3(S^3)$).

Thus, the relation between an exotic $\mathbb{R}^4$ (of Bizaca as constructed from the failure of the smooth h-cobordism theorem) and codimension-one foliation of the S^3 emerges. Two non-diffeomorphic exotic $\mathbb{R}^4$ imply non-cobordant codimension-one foliations of the 3-sphere described by the Godbillon-Vey class in $H^3(S^3,\mathbb{R})$ (proportional to the surface of the polygon). This relation is very strict, i.e. if we change the Casson handle, then we must change the polygon. But that changes the foliation and vice verse. Finally, we obtain the result:

The exotic $\mathbb{R}^4$ (of Bizaca) is determined by the codimension-1 foliations with non-vanishing Godbillon-Vey class in $H^3(S^3,\mathbb{R}^3)$ of a 3-sphere seen as submanifold $S^3 \subset \mathbb{R}^4$. We say: the exoticness is localized at a 3-sphere inside the small exotic $\mathbb{R}^4$.

In the particular case of integral $H^3(S^3,\mathbb{Z})$ one yields the relation of exotic $\mathbb{R}^4_k$, $k[\,] \in H^3(S^3,\mathbb{Z})$, $k \in \mathbb{Z}$ with the WZ term of the k WZW model on $SU(2)$. This is because the integer classes in $H^3(S^3,\mathbb{Z})$ are of special character. Topologically, this case refers to flat $PSL(2,\mathbb{R})-$bundles over the space $(S^2 \setminus \{\text{k punctures}\}) \times S^1$ where the gluing of k solid tori produces a 3-sphere (so-called Heegard decomposition). Then, one obtains the relation [Asselmeyer-Maluga & Król (2009a)]:

$$\frac{1}{(4\pi)^2}\langle GV(\mathcal{F}),[S^3]\rangle = \frac{1}{(4\pi)^2}\int\limits_{S^3} GV(\mathcal{F}) = \pm(2-k) \tag{1}$$

in dependence on the orientation of the fundamental class $[S^3]$. We can interpret the Godbillon-Vey invariant as WZ term. For that purpose, we use the group structure $SU(2) = S^3$ of the 3-sphere S^3 and identify $SU(2) = S^3$. Let $g \in SU(2)$ be a unitary matrix with

$\det g = -1$. The left invariant 1-form $g^{-1}dg$ generates locally the cotangent space connected to the unit. The forms $\omega_k = Tr((g^{-1}dg)^k)$ are complex $k-$forms generating the deRham cohomology of the Lie group. The cohomology classes of the forms ω_1, ω_2 vanish and $\omega_3 \in H^3(SU(2), \mathbb{R})$ generates the cohomology group. Then, we obtain as the value for the integral of the generator

$$\frac{1}{8\pi^2} \int_{S^3=SU(2)} \omega_3 = 1\,.$$

This integral can be interpreted as winding number of g. Now, we consider a smooth map $G : S^3 \to SU(2)$ with 3-form $\Omega_3 = Tr((G^{-1}dG)^3)$ so that the integral

$$\frac{1}{8\pi^2} \int_{S^3=SU(2)} \Omega_3 = \frac{1}{8\pi^2} \int_{S^3} Tr((G^{-1}dG)^3) \in \mathbb{Z}$$

is the winding number of G. Every Godbillon-Vey class with integer value like (1) is generated by a 3-form Ω_3. Therefore, the Godbillon-Vey class is the WZ term of the $SU(2)_k$. Thus, we obtain the relation:

The structure of exotic $\mathbb{R}^4_k$'s, $k \in \mathbb{Z}$ from the radial family determines the WZ term of the k WZW model on $SU(2)$.

This WZ term enables one for the cancellation of the quantum anomaly due to the conformal invariance of the classical σ-model on $SU(2)$. Thus, we have a method of including this cancellation term from smooth 4-geometry: when a smoothness of the ambient 4-space, in which S^3 is placed as a part of the boundary of the cork, is this of exotic $\mathbb{R}^4_k$, then the WZ term of the classical σ-model with target $S^3 = SU(2)$, i.e. $SU(2)_k$ WZW, is precisely generated by this 4-smoothness. As the conclusion, we have the important correlation:

The change of smoothnes of exotic $\mathbb{R}^4_k$ to exotic $\mathbb{R}^4_l$, $k, l \in \mathbb{Z}$ both from the radial family, corresponds to the change of the level k of the WZW model on $SU(2)$, i.e. k WZW $\to$l WZW.

Let us consider now the end of the exotic $\mathbb{R}^4_k$ i.e. $S^3 \times \mathbb{R}$. This end cannot be standard smooth and it is in fact fake smooth $S^3 \times_{\Theta_k} \mathbb{R}$, Freedman (1979). Given the connection of $\mathbb{R}^4_k$ with the WZ term as above, we have determined the "quantized" geometry of $SU(2)_k \times \mathbb{R}$ as corresponding to the exotic geometry of the end of $\mathbb{R}^4_k$. The appearance of the $SU(2)_k \times \mathbb{R}$ is a source for various further constructions. In particular, we will see that gravitational effects of $\mathbb{R}^4_k$ on the quantum level are determined via string theory where one replaces consistently flat $\mathbb{R}^4$ part of the background by curved 4D $SU(2)_k \times \mathbb{R}$.

3. 10d string theory and 4d-smoothness

Let us, following Asselmeyer-Maluga & Król (2011a) (see also Król (2011a;b)), begin with a charged quantum particle, say e, moving through non-flat gravitational background, i.e. smooth 4-spacetime manifold. The amount of gravity due to the curvature of this background affects the particle trajectory as predicted by GR. There should exist, however, a high energy limit where gravity contained in this geometrical background becomes quantum rather than classical and the particle may not be described by perturbative field theory any longer. This rather natural, from the point of view of physics, scenario requires, however, quantum gravity calculations which is not in reach in dimension 4. Moreover, mathematics underlying classical

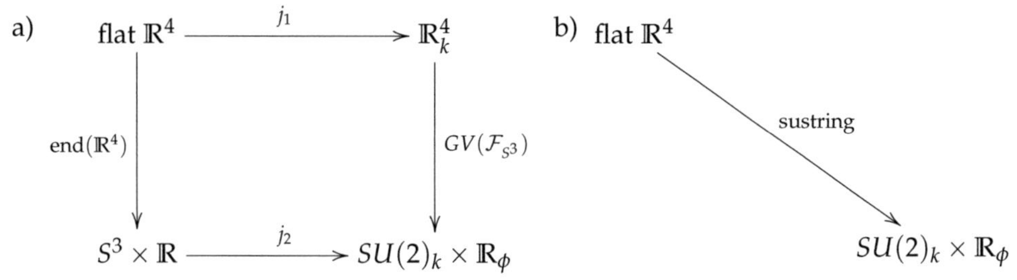

Fig. 1. a) j_1 is the change of the standard smooth $\mathbb{R}^4$ to the exotic $\mathbb{R}^4_k$, end$(\mathbb{R}^4)$ assigns the standard end to $\mathbb{R}^4$, $GV(\mathcal{F}_{S^3})$ generates the $\mathrm{WZ_k}$-term from exotic $\mathbb{R}^4_k$ via GV invariant of the codim.-1 foliation of S^3. b) The change of string backgrounds s.t. flat $\mathbb{R}^4$ part is replaced by the linear dilaton background $SU(2)_k \times \mathbb{R}_\phi$

gravity is of (pseudo-)Riemannian smooth geometry and should change to a new 'geometry, or to, unknown at all, mathematics, when the transformation of GR to QG is performed.

As discussed already in the Introduction, it was proposed in Asselmeyer-Maluga & Król (2010; 2011c;d; 2012); Król (2011a;b) that 4-dimensional effects of string theory should be seen via its connections with exotic smoothness of topologically trivial $\mathbb{R}^4$. Several results were derived as if superstring theory were formulated on backgrounds which contain 4-dimensional part which is exotic $\mathbb{R}^4$ rather than standard smooth $\mathbb{R}^4$. This serves as a new window to 4-dimensional physics. The argumentation dealt with exact string backgrounds in any order of α'. The existence of such backgrounds is rather exceptional in superstring theory (see e.g. Orlando (2006; 2007)) and this always indicates important and exactly calculable effects of the theory. This is precisely the tool which we want to apply to the above stated problems, i.e. the description of both, 4D QG effects due to gravity present in background spacetime, and mathematics behind the shift GR $\to$ QG in 4D.

Exotic $\mathbb{R}^4_k$ is a smooth Riemannian manifold, however, its structure essentially deals with non-commutative geometry and quantization Asselmeyer-Maluga & Król (2011b). The connection with string exact backgrounds was also recognized in Asselmeyer-Maluga & Król (2010; 2011c). Thus, under the topological assumptions discussed in Sec. 2, the following correspondence emerges:

> The change of the smoothness from the standard $\mathbb{R}^4$ to exotic $\mathbb{R}^4_k$, corresponds to the change of exact string backgrounds from $\mathbb{R}^4 \times K^6$ to $SU(2)_k \times \mathbb{R}_\phi \times K^6$.

Let us note that only because of the *exotic* smooth structure of $\mathbb{R}^4$, the link to string backgrounds exists. If smoothness of $\mathbb{R}^4$ were standard, only separated regimes of 4-geometry (GR) and superstrings (QG) would appear. In superstring theory one understands fairy well how to change the exact background containing flat $\mathbb{R}^4$ to this with curved 4-dimensional part: $\mathbb{R}^4 \times K^6 \to SU(2)_k \times \mathbb{R}_\phi \times K^6$. This requires supersymmetry in 10 dimensions. The presence of supersymmetry is, however, just a technical mean allowing for the consistent shifts between the backgrounds and performing the QG calculations effectively.

3.1 The magnetic deformation of 4D part of the string background

To be specific let us consider the $SO(3)_{k/2} \times \mathbb{R}_\phi$ as the 4D part of the 10D string background which replaces the flat $\mathbb{R}^4$ part. This $SO(3)_{k/2} \times \mathbb{R}_\phi$ geometry is the result of the projection from $SU(2)_k \times \mathbb{R}_\phi$, so k is even. Here $SU(2)_k$ is the affine Kac-Moody algebra at level k and $\mathbb{R}_\phi$ the linear dilaton, both appearing in the exact string background realized by the superconformal 2D field theory (see, e.g. Kiritsis & Kounnas (1995b)). The deformation of such curved 4D part of the background will be performed in heterotic superstring theory in the language of σ-model. The deformations will correspond to the introducing almost constant magnetic field H and its gravitational backreaction on the 4D curved part of the background. First let us describe undeformed theory. The action for heterotic σ-model in this $SO(3)_{k/2} \times \mathbb{R}_\phi$ background is:

$$S_4 = \frac{k}{4} \mathrm{I}_{SO(3)}(\alpha, \beta, \gamma) + \frac{1}{2\pi} \int d^2z \left[\partial x^0 \bar{\partial} x^0 + \psi^0 \partial \psi^0 + \sum_{a=1}^{3} \psi^a \partial \psi^a \right] + \frac{Q}{4\pi} \int \sqrt{g} R^{(2)} x^0 \quad (2)$$

where $\mathrm{I}_{SO(3)}(\alpha, \beta, \gamma) = \frac{1}{2\pi} \int d^2z \left[\partial\alpha\bar{\partial}\alpha + \partial\beta\bar{\partial}\beta + \partial\gamma\bar{\partial}\gamma + 2\cos\beta\partial\alpha\partial\gamma \right]$ in Euler angles of $SU(2) = S^3$, $R^{(2)}$ is the 2D worldsheet curvature, g is the determinant of the target metric and Q is the dilaton charge with x^0 the coordinate of $\mathbb{R}_\phi$. The bosonic σ-model action reads in general:

$$S = \frac{1}{2\pi} \int d^2z (G_{\mu\nu} + B_{\mu\nu}) \partial x^\mu \bar{\partial} x^\nu + \frac{1}{4\pi} \int \sqrt{g} R^{(2)} \Phi(x) \quad (3)$$

so comparing with (2) gives the non-zero background fields as:

$$\begin{gathered} G_{00} = 1,\ G_{\alpha\alpha} = G_{\beta\beta} = G_{\gamma\gamma} = \tfrac{k}{4},\ G_{\alpha\gamma} = \tfrac{k}{4} \cos\beta \\ B_{\alpha\gamma} = \tfrac{k}{4} \cos\beta,\ \Phi = Qx^0 = \frac{x^0}{\sqrt{k+2}}\,. \end{gathered} \quad (4)$$

One can decompose (see e.g. Prezas & Sfetsos (2008)) the supersymmetric WZW model into the bosonic $SU(2)_{k-2}$ with affine currents J^i and three free fermions ψ^a, $a = 1, 2, 3$ in the adjoint representation of $SU(2)$. As the result the supersymmetric $\mathcal{N} = 1$ affine currents are $\mathcal{J}^a = J^a - \frac{i}{2} \epsilon^{abc} \psi^b \psi^c$. After introducing the complex fermions combination $\psi^\pm = \frac{1}{\sqrt{2}} (\psi^1 \pm i\psi^2)$ and the corresponding change of the affine bosonic currents $J^\pm = J^1 \pm iJ^2$, the supersymmetric affine currents read:

$$\mathcal{J}^3 = J^3 + \psi^+\psi^-,\ \mathcal{J}^\pm = J^\pm \pm \sqrt{2}\psi^3\psi^\pm \quad (5)$$

Let us redefine the indices in the fermion fields as: $+ \to 1$, $- \to 2$, then $\mathcal{J}^3 = J^3 + \psi^1\psi^2$.

From the point of view of the σ-model, the vertex for the magnetic field H on 4-dimensional $\mathbb{R}_\phi \times SU(2)_k$ part of the background is the exact marginal operator given by $V_m = H(J^3 + \psi^1\psi^2)\bar{J}^a$. Similarly, the vertex for the corresponding gravitational part is $V_{gr} = \mathcal{R}(J^3 + \psi^1\psi^2)\bar{J}^3$, and represents truly marginal deformations too.

The shape of these operators follow from the fact that, in general, the marginal deformations of the WZW model can be constructed as bilinears in the currents $J, \bar{J}$ of the model [Orlando (2007)]:

$$\mathcal{O}(z, \bar{z}) = \sum_{i,j} c_{ij} J^i(z) \bar{J}^j(\bar{z}) \quad (6)$$

where $J^i, \bar{J^j}$ are left and right-moving affine currents respectively, Orlando (2007).

Here, following Kiritsis & Kounnas (1995b), we consider covariantly constant magnetic field $H_i^a = \epsilon^{ijk} F_{jk}^a$ and constant curvature $\mathcal{R}^{il} = \epsilon^{ijk} \epsilon^{lmn} \mathcal{R}_{jmkn}$ in the 4-dimensional background as above of closed superstring theory. When this chromo-magnetic field is in the $\mu = 3$ direction the following deformation is proportional to $(J^3 + \psi^1\psi^2)\bar{J}$ and the right moving current $\bar{J}$ is normalized as $< \bar{J}(1)\bar{J}(0) >= k_g/2$. Rewriting the currents in the Euler angles, i.e. $J^3 = k(\partial\gamma + \cos\beta\partial\alpha)$, $\bar{J}^3 = k(\bar{\partial}\alpha + \cos\beta\bar{\partial}\gamma)$, we obtain for the perturbation of the (heterotic) action in (2), the following expression:

$$\delta S_4 = \frac{\sqrt{kk_g}H}{2\pi}\int d^2z(\partial\gamma + \cos\beta\partial\alpha)\bar{J}\,. \tag{7}$$

The new σ-model with the action $S_4 + \delta S_4$ is again conformally invariant with all orders in α' since:

$S_4 + \delta S_4 = \frac{k}{4}I_{SO(3)}(\alpha,\beta,\gamma) + \delta S_4 + \frac{k_g}{4\pi}\int d^2z\partial\phi\bar{\partial}\phi = \frac{k}{4}I_{SO(3)}(\alpha,\beta,\gamma + 2\sqrt{\frac{k_g}{k}}H\phi) + \frac{k_g(1-2H^2)}{4\pi}\int d^2z\partial\phi\bar{\partial}\phi$. This shows that, in fact the magnetic deformation is exactly marginal. Here we have chosen for the currents J and $\bar{J}$, $\partial\phi$ and $\bar{\partial}\phi$ correspondingly, as their bosonizations.

The background corresponding to the perturbation (7) is determined by background fields, i.e. a graviton $G_{\mu\nu}$, gauge fields $F^a_{\mu\nu}$, an antisymmetric field (three form) $H_{\mu\nu\rho}$ and a dilaton Φ, which, in turn, are solutions to the following equations of motion:

$$\begin{gathered}\frac{3}{2}\left[4(\nabla\Phi)^2 - \frac{10}{3}\Box\Phi - \frac{2}{3}R + \frac{1}{12g^2}F^a_{\mu\nu}F^{a,\mu\nu}\right] = C\\ R_{\mu\nu} - \frac{1}{4}H^2_{\mu\nu} - \frac{1}{2g^2}F^a_{\mu\rho}F^{a\rho}_{\nu} + 2\nabla_\mu\nabla_\nu\Phi = 0\\ \nabla^\mu\left[e^{-2\Phi}H_{\mu\nu\rho}\right] = 0\\ \nabla^\nu\left[e^{-2\Phi}F^a_{\mu\nu}\right] - \frac{1}{2}F^{a,\nu\rho}H_{\mu\nu\rho}e^{-2\Phi} = 0\end{gathered} \tag{8}$$

These are derived from the variations of the following effective 4-dimensional gauge theory action:

$$S = \int d^4x\sqrt{G}e^{-2\Phi}\left[R + 4(\nabla\Phi)^2 - \frac{1}{12}H^2 - \frac{1}{4g^2}F^a_{\mu\nu}F^{a,\mu\nu} + \frac{C}{3}\right] \tag{9}$$

where C is the l.h.s. of the first equation in (8). Here $g_{str} = 1$, the gauge coupling $g^2 = 2/k_g$, $F^a_{\mu\nu} = \partial_\mu A_\nu - \partial_\nu A_\mu + f^{abc}A^b_\mu A^c_\nu$, $H_{\mu\nu\rho} = \partial_\mu B_{\nu\rho} - \frac{1}{2g^2}\left[A^a_\mu F^a_{\nu\rho} - \frac{1}{3}f^{abc}A^a_\mu A^b_\nu A^c_\rho\right] +$ permutations. f^{abc} are structure constants of the gauge group and A^a_μ is the effective gauge field. One can observe that the term in the square bracket in $H_{\mu\nu\rho}$ is the Chern-Simons term for the gauge potential A^a_μ.

Now, the background complying with these equations and which respects the deformation (7), reads:

$$\begin{gathered}G_{00} = 1,\ G_{\beta\beta} = \frac{k}{4},\ G_{\alpha\gamma} = \frac{k}{4}(1 - 2H^2)\cos\beta\\ G_{\alpha\alpha} = \frac{k}{4}(1 - 2H^2\cos^2\beta),\ G_{\gamma\gamma} = \frac{k}{4}(1 - 2H^2),\ B_{\alpha\gamma} = \frac{k}{4}\cos\beta\\ A_\alpha = g\sqrt{k}H\cos\beta,\ A_\gamma = g\sqrt{k}H,\ \Phi = \frac{x^0}{\sqrt{k+2}}\,.\end{gathered} \tag{10}$$

where H is the magnetic field as in (7).

Similarly, when gravitational marginal deformations as in the vertex $V_{gr} = \mathcal{R}(J^3 + \psi^1\psi^2)\bar{J}^3$ are included, where $\mathcal{R}$ is the curvature parameter of the deformation, one can derive corresponding exact background of string theory via σ-model calculations, Hassan & Sen (1993); Kiritsis & Kounnas (1995b). Again, the fields in this background which solve the effective field theory equations (8), are [Kiritsis & Kounnas (1995b)]:

$$
\begin{aligned}
G_{00} &= 1,\ G_{\beta\beta} = \tfrac{k}{4} \\
G_{\alpha\alpha} &= \tfrac{k}{4}\frac{(\lambda^2+1)^2-(8H^2\lambda^2+(\lambda^2-1)^2)\cos^2\beta)}{(\lambda^2+1+(\lambda^2-1)\cos\beta)^2} \\
G_{\gamma\gamma} &= \tfrac{k}{4}\frac{(\lambda^2+1)^2-(8H^2\lambda^2-(\lambda^2-1)^2)\cos^2\beta)}{(\lambda^2+1+(\lambda^2-1)\cos\beta)^2} \\
G_{\alpha\gamma} &= \tfrac{k}{4}\frac{4\lambda^2(1-2H^2)\cos\beta+(\lambda^4-1)\sin^2\beta}{(\lambda^2+1+(\lambda^2-1)\cos\beta)^2} \\
B_{\alpha\gamma} &= \tfrac{k}{4}\frac{\lambda^2-1+(\lambda^2+1)\cos\beta}{(\lambda^2+1+(\lambda^2-1)\cos\beta)^2} \\
A_\alpha &= 2g\sqrt{k}\frac{H\lambda\cos\beta}{(\lambda^2+1+(\lambda^2-1)\cos\beta)^2} \\
A_\gamma &= 2g\sqrt{k}\frac{H\lambda}{(\lambda^2+1+(\lambda^2-1)\cos\beta)^2} \\
\Phi &= \tfrac{t}{\sqrt{k+2}} - \tfrac{1}{2}\log\left[\lambda+\tfrac{1}{\lambda}+(\lambda-\tfrac{1}{\lambda})\cos\beta\right]
\end{aligned}
\tag{11}
$$

The dependence on λ shows the existence of gravitational backreaction which was absent in the purely magnetic deformed background (10).

3.2 Field theory vs. string theory spectra of charged particles in standard 4-space

In the case of field theory in 4 dimensions we introduce the magnetic field on S^3 which agrees with the magnetic part of the string background (10) as:

$$A_\alpha = H\cos\beta,\ A_\beta = 0,\ A_\gamma = H. \tag{12}$$

The Hamiltonian for a particle with electric charge e moving on S^3, is

$$\overline{\mathbf{H}} = \frac{1}{\sqrt{\det G}}(\partial_\mu - ieA_\mu)\sqrt{\det G}G^{\mu\nu}(\partial_\nu - ieA_\nu). \tag{13}$$

where we assume at the beginning that $G_{\mu\nu}$ is standard metric on S^3.

The energy spectrum for $\overline{\mathbf{H}}$ is then given by:

$$\Delta E_{j,m} = \frac{1}{R^2}\left[j(j+1) - m^2 + (eH-m)^2\right] \tag{14}$$

where R is the radius of S^3, $j \in \mathbb{Z}$ and $-j \le m \le j$, as is the case for $SO(3)$. In the flat limit we retrieve the Landau spectrum in 3-dimensional space of spinless particles:

$$\Delta E_{n,p_3} = e\tilde{H}(2n+1) + p_3^2 + \mathcal{O}(R^{-1}) \tag{15}$$

where magnetic field is pointing into 3-rd direction and the re-scaling of eH is performed as $eH = e\tilde{H} + \kappa R + \mathcal{O}(1)$, $m = e\tilde{H}R^2 + (p_3+\kappa) + \mathcal{O}(1)$. This follows from rewriting the spectrum (14) as $\Delta E_{n,m} = \frac{1}{R^2}\left[n(n+1) + |m|(2n+1)\right] + \left(\frac{eH-m^2}{R}\right)^2$ by introducing new parameter n: $j = |m| + n$ for $|m|, n \in \mathbb{N}$.

Let us, again following Kiritsis & Kounnas (1995b), calculate the spectrum in the case of full exact string background (10) as our starting point. One takes the metric components from the background (10) and derive the eigenvalues of the Hamiltonian (14). The result is [Kiritsis & Kounnas (1995b)]:

$$\Delta E_{j,m} = \frac{1}{R^2}\left[j(j+1) - m^2 + \frac{(eHR-m)^2}{(1-2H^2)}\right] . \tag{16}$$

Again, introducing $n \in \mathbb{N}$ by $j = |m| + n$, $|m| = 0, 1/2, 1, \ldots$ we can rewrite the spectrum (16) as:

$$\Delta E_{n,m} = \frac{1}{R^2}[n(n+1) + |m|(2n+1)] + \left(\frac{eHR-m}{R\sqrt{1-2H^2}}\right)^2 \tag{17}$$

which is the energy spectrum containing the corrections due to H field appearing in the string exact background (10), but the Hamiltonian (13) is field theoretic 4-dimensional one.

One can also calculate the exact string spectrum of energy in this exact background (see Kiritsis & Kounnas (1995a;b)) and when compared with (17) gives rise to the following dictionary rules enabling passing between the spectra:

$$\begin{gathered} R^2 \to k+2\,,\; m \to \mathcal{Q} + J^3\,,\; e \to \sqrt{\tfrac{2}{k_g}}\,\overline{\mathcal{Q}} \\ H \to \frac{F}{\sqrt{2}(1+\sqrt{1+F^2})} = \frac{1}{2\sqrt{2}}\left[F - \frac{F^3}{4} + \mathcal{O}(F^5)\right] . \end{gathered} \tag{18}$$

Here $F^2 = \left\langle F^a_{\mu\nu} F^{\mu\nu}_a \right\rangle$ is the integrated (square of) field strength where $H^a_i = \epsilon^{ijk} F^a_{jk}$ as before.

For a particle with spin S setting $S = Q$ the following modification of the spectrum appear due to the above rules [Kiritsis & Kounnas (1995b)]:

$$\Delta E_{j,m,S} = \frac{1}{k+2}\left[j(j+1) - (m+S)^2 + \frac{(eHR-m-S)^2}{(1-2H^2)}\right] . \tag{19}$$

Next step is the inclusion of gravitational backreactions. One begins with the string background (11) and compute again the eigenvalues of (13). The result for scalar particles is [Kiritsis & Kounnas (1995b)]:

$$\Delta E_{j,m,\overline{m}} = \frac{1}{R^2}\left[j(j+1) - m^2 + \frac{(2ReH - (\lambda + \frac{1}{\lambda})m - (\lambda - \frac{1}{\lambda})\overline{m})^2}{4(1-2H^2)}\right] \tag{20}$$

where now $-j \leq m, \overline{m} \leq j$. Again, comparing with exact string spectra for even k we have the corresponding dictionary rules in the case where gravity backreactions are included:

$$\begin{gathered} R^2 \to k+2\,,\; m \to \mathcal{Q} + J^3\,,\; e \to \sqrt{\tfrac{2}{k_g}}\,\overline{\mathcal{P}}\,,\; \overline{m} \to \overline{J}^3 \\ H^2 \to \frac{1}{2}\frac{F^2}{F^2 + 2(1+\sqrt{1+F^2+\mathcal{R}^2})}\,,\; \lambda^2 = \frac{1+\sqrt{1+F^2+\mathcal{R}^2}+\mathcal{R}}{1+\sqrt{1+F^2+\mathcal{R}^2}-\mathcal{R}} \end{gathered} \tag{21}$$

where $\mathcal{R}^2 = \langle R_{\mu\nu\rho\sigma} R^{\mu\nu\rho\sigma} \rangle$ is the integrated squared scalar curvature and $R_{\mu\nu\rho\sigma}$ is the Riemann tensor of the „squashed" $SU(2) = S^3$ in the deformed background.

3.3 The exotic 4D interpretation of string calculations

Given the dictionary (21), we can rewrite (20) in a way where the dependance on the even level k is written explicitly:

$$\Delta E^k_{j,m,\overline{m}} = \frac{1}{k+2}[j(j+1)-m^2] + \frac{(2\sqrt{k+2}eH - (\lambda+\frac{1}{\lambda})m - (\lambda-\frac{1}{\lambda})\sqrt{(1+2/k)}\overline{m})^2}{4(k+2)(1-2H^2)}. \quad (22)$$

Thus this is the 4D spectrum of a scalar particle with charge e which is modified by the magnetic field H and its gravitational backreaction λ (20). The spectrum depends on $k = 2p$ which indicates the relevance of the stringy regime. One can interpret this dependance on k as the result of exotic $\mathbb{R}^4_k$ geometry of a 4-region where the particle travells. However, this 4-geometry, in the QG limit of string theory, generates the quantum gravity effects in 4D.

In deriving the spectrum (22) we commence with the flat standard smooth $\mathbb{R}^4$ which is a part of the exact string background. Then we switched to another exact string background where the 4D part is now $SU(2)_k \times \mathbb{R}_\phi$. This new 4D part ceases to be flat. Its curvature has defined gravitational meaning in superstring theory such that the QG calculations are possible. The effects are derived in the regime of QG, i.e. heterotic string theory. The same deformed spectrum could be obtained, in principle, via including magnetic field $\tilde{H}$ and its gravitational backreaction on exotic smooth $\mathbb{R}^4_k$ where modified metric $\tilde{G}_{\mu\nu}$ emerges. These fields, however, are not explicitly specified but still the effects in QG regime are derived from string theory as above. Such an approach serves as a way of quantization of gravity while on exotic $\mathbb{R}^4$. The relations between various ingredients appearing here are presented in Fig. 2.

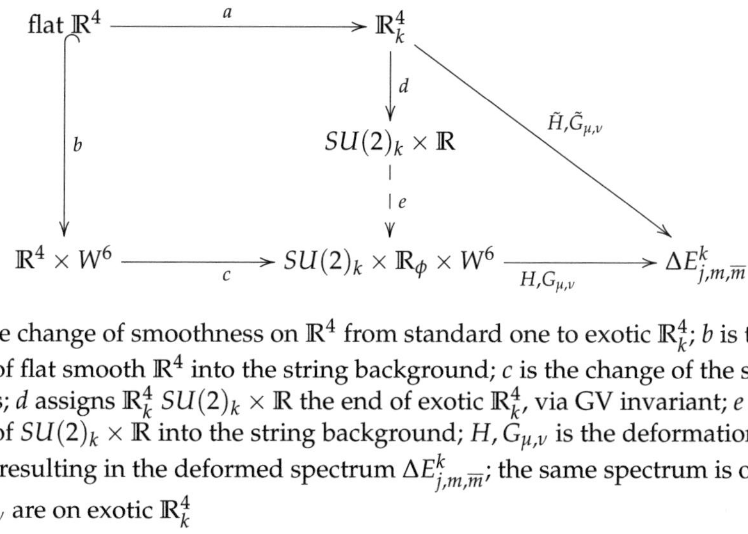

Fig. 2. a is the change of smoothness on $\mathbb{R}^4$ from standard one to exotic $\mathbb{R}^4_k$; b is the embedding of flat smooth $\mathbb{R}^4$ into the string background; c is the change of the string backgrounds; d assigns $\mathbb{R}^4_k$ $SU(2)_k \times \mathbb{R}$ the end of exotic $\mathbb{R}^4_k$, via GV invariant; e is the embedding of $SU(2)_k \times \mathbb{R}$ into the string background; $H, G_{\mu,\nu}$ is the deformation of the CFT background resulting in the deformed spectrum $\Delta E^k_{j,m,\overline{m}}$; the same spectrum is obtained when $\tilde{H}, \tilde{G}_{\mu,\nu}$ are on exotic $\mathbb{R}^4_k$

Let us turn to the appearance of the mass gap in the spectrum when the theory is formulated on exotic $\mathbb{R}^4_k$ rather than standard smooth $\mathbb{R}^4$. In field theory a dilaton Φ couples to a massless bosonic field T in a universal fashion:

$$S[\Phi, T] = \int e^{-2\Phi} \partial_M T \partial^M T.$$

One can introduce a new field $U = e^{-\Phi}T$ hence the above action becomes:

$$S[\Phi, U] = \int \partial_M U \partial^M U + [\partial^2\Phi - \partial_M \Phi \partial^M \Phi] U.$$

Thus, for a linear dilaton $\Phi = q_M X^M$ the field U gets a mass square $M^2 = q_M q^M$ for q_M spacelike. This way the massless boson T is mapped to the boson U with the mass M. However, this mechanism does not work in the case of massless free fermions. In four dimensional spacetime the chiral fermion ψ can be coupled to an antisymmetric tensor $H_{\mu\nu\rho}$ as follows:

$$S[\psi, H] = \int \overline{\psi}\gamma^\mu \left[\overleftrightarrow{\partial_\mu} + H_\mu\right] \psi$$

where $H_\mu = \varepsilon^{\mu\nu\rho\sigma} H_{\nu\rho\sigma}$ is the dual of the antisymmetric tensor $H_{\nu\rho\sigma}$. If one can embed this system into a string background with the fields: Φ and H_{MNP}, then using one-loop string equations:

$$\begin{aligned} R_{MN} &= -2\nabla_M \nabla_N \Phi + \tfrac{1}{4} H_{MPR} H_N{}^{PR}, \\ \nabla_L \left(e^{-2\Phi} H^L_{MN}\right) &= 0, \\ \nabla^2 \Phi - 2\left(\nabla\Phi\right)^2 &= -\tfrac{1}{12} H^2, \end{aligned} \tag{23}$$

one gets for the linear dilaton $\Phi = q_M X^M$ the following relation:

$$q_M q^M = \frac{1}{6} H^2$$

and the scalar curvature R is:

$$R = \frac{3}{2} q_M q^M .$$

If non-vanishing components of q_M and H_{MNP} are in four dimensional space, one obtains that:

$$q^\mu \sim \varepsilon^{\mu\nu\rho\sigma} H_{\nu\rho\sigma}.$$

Thus, the Dirac operator acquires a mass gap proportional to $q_\mu q^\mu$.

The problem of embedding a four dimensional fermion system in the exact string background was considered in Kiritsis & Kounnas (1995c). In the case when four dimensional space is represented by the $\mathbb{R}_\phi \times SU(2)_k$ part of the string background, than the linear dilaton is $\Phi = QX^0$ and Q is given by the level k of the WZW model on $SU\,(2)$ as $Q = (k+2)^{-1/2}$ so that the CFT has the same central charge as flat space. Hence, the massless bosons acquire the mass gap $\Delta M^2 = \mu^2 = (k+2)^{-1}$.

That way we arrive at the important feature of the theory when on exotic $\mathbb{R}^4_k$:

The theory predicting the energy spectra of charged particles as in (22) in the flat smooth 4D limit, $k \to \infty$ does not show the existence of the mass gap in the energy spectra. However, in the exotic $\mathbb{R}^4_k$ limit the theory acquires the mass gap $\mu^2 \sim \frac{1}{(k+2)}$.

The mass gap which appears here in 4D theory is the result of QG computations i.e. those on linear dilaton background in superstring theory. Such overlapping QG with field theory in 4D is a special new feature of the approach via exotic open 4-smooth spaces which bridges 10D superstring and 4D matter fields. We will see in the next section that there are also bottom–up arguments where exotic $\mathbb{R}^4$'s emerge from the regime of low energy effective states of condensed matter. The latter means that gravity is present in the description of the effective entangled matter, since $\mathbb{R}^4_k$ is not flat and Einstein equations can be written on these 4-manifolds.

4. Quantum effective spin matter and exotic $\mathbb{R}^4$ – the Kondo effect

Gravitational interaction is very exceptional among all interactions in Nature. On the one hand gravity is the geometry of spacetime on which fields propagate and interactions take place. On the other hand, gravity couples with any kind of energy and matter. Further, it is the only interaction which restrains quantization.

Based on the entanglement of ideas presented so far, we want to argue that gravity is present in some states of magnetic effective quantum matter in a *nonstandard* way. The latter means that some states of spin matter, already at low temperatures, are coupled with 4D gravity via special 4-geometry *directly*, rather than, by energy-momentum tensor. This coupling can be extended over quantum regime of gravity, at least in some cases, and relates effective rather than fundamental fields and particles from SM. The coupling is understood as the presence of a non-flat 4-geometry which becomes dominating in some limits. The special 4-geometry is, again, exotic smoothness of Euclidean 4-space $\mathbb{R}^4$, thus becoming a guiding principle for presented approach to QG. The presence of gravity in the description of nonperturbative, strongly entangled states of 4D matter field is not a big surprise, as recent vital activity on the methods of AdS/CFT correspondence shows. However, our approach is different and makes use of inherently 4-dimensional new geometrical findings, which, at this stage of development, do not refer to AdS/CFT techniques (cf. Król (2005)).

In the thirties of the last century strange behaviour of conducting electrons occurring in some metallic alloys was observed. Namely the resistivity $\rho(T)$ in these alloys in the presence of magnetic spin s impurities, growth substantially when the temperature is lowering below the critical temperature T_K called the Kondo temperature. T_K is as low as a few K.

Kondo proposed in 1964 a simple phenomenological Hamiltonian Affleck (1995):

$$H = \sum_{\vec{k},\alpha} \psi^{+\alpha}_{\vec{k}} \psi_{\vec{k}\alpha} \epsilon(k) + \lambda \vec{S} \cdot \sum_{\vec{k},\vec{k}'} \psi^{+}_{\vec{k}} \frac{\vec{\sigma}}{2} \psi_{\vec{k}'} . \tag{24}$$

explaining the growth of the resistivity $\rho(T)$. Here ψ is the annihilation operator for the conduction electron of spin α and momentum $\vec{k}$, the antiferromagnetic interaction term is that between spin s impurity $\vec{S}$ with spins of conducting electrons, at $\vec{x} = 0$; $\vec{\sigma}$ is the vector of Pauli matrices. From this Hamiltonian one can derive, in the Born approximation, that $\rho(T) \sim \left[\lambda + \nu\lambda^2 \ln \frac{D}{T} + ...\right]^2$ where D is the 'width of the band' parameter and the second term is divergent in $T = 0$. Thus, this divergence explains the growth of the resistivity. The Hamiltonian (24) can be also derived from the more microscopic Anderson model Anderson (1961). The Kondo antiferromagnetic coupling appears as the tunnelling of electrons thus screening the spin impurity (see eg. Potok et al. (2007)).

The exact low T behavior was proposed by Affleck (1995); Affleck & Ludwig (1991; 1993; 1994) and Potok et al. (2007) by the use of boundary conformal field theory (BCFT). This insightful use of the CFT methods makes it possible to work out the connection with smooth 4-geometry.

Let us see in brief how the structure of the $SU(2)_k$ WZW model is well suited to the description of the k-channel Kondo effect. Recall that Kac-Moody algebra $SU(2)_k$ is spanned on 3-components currents $\vec{\mathcal{J}}_n, n = ... - 2, -1, 0, 1, 2, ...$:

$$[\mathcal{J}_n^a, \mathcal{J}_m^b]_k = i\epsilon^{abc}\mathcal{J}_{n+m}^c + \frac{1}{2}kn\delta^{ab}\delta_{n,-m} \,. \tag{25}$$

Next, we decompose the currents $\vec{\mathcal{J}}_n$ as $\vec{\mathcal{J}}_n = \vec{J}_n + \vec{S}$ such that $\vec{J}_n$ obey the same Kac-Moody algebra, i.e. $[J_n^a, J_m^b]_k = i\epsilon^{abc}J_{n+m}^c + \frac{1}{2}kn\delta^{ab}\delta_{n,-m}$ and usual relations for $\vec{S}$, i.e. $[S^a, S^b] = i\epsilon^{abc}S^c, [S^a, J_n^b] = 0$. From the point of view of field theories describing the interacting currents with spins, $\vec{\mathcal{J}}_n$ corresponds to the effective infrared fixed point of the theory of interacting spins $\vec{S}$ with $\vec{J}_n$where the coupling constant λ is taken as $\frac{2}{3}$ for $k = 1$. The interacting Hamiltonian of the theory, for $k = 1$, reads:

$$H_s = c\left(\frac{1}{3}\sum_{-\infty}^{+\infty}\vec{J}_{-n}\cdot\vec{J}_n + \lambda\sum_{-\infty}^{+\infty}\vec{J}_n\cdot\vec{S}\right) . \tag{26}$$

For $\lambda = \frac{2}{3}$, one completes the square and the algebra (25) for the currents $\vec{\mathcal{J}}_n$ follows. Then, the new Hamiltonian, where $\vec{S}$ is now effectively absent (still for $k = 1$), is given by $H = c'\sum_{-\infty}^{+\infty}\left(\vec{\mathcal{J}}_{-n}\cdot\vec{\mathcal{J}}_n - \frac{3}{4}\right)$ (c, c' are some constants).

A similar procedure holds for arbitrary integer k where the spin part of the Hamiltonian reads: $H_{s,k} = \frac{1}{2\pi(k+2)}\vec{J}^2 + \lambda\vec{J}\cdot\vec{S}\delta(x)$ and the infrared effective fixed point is now reached for $k = \frac{2}{2+k}$. The spins $\vec{S}$ reappear as the boundary conditions in the boundary CFT represented by the WZW model on $SU(2)$. This model defines the Verlinde fusion rules and is determined by these. The following *fusion rules hypothesis*, was proposed by Affleck (1995), which explains the creation and nature of the multichannel Kondo states:

The infrared fixed point, in the k-channel spin-s Kondo problem, is given by fusion with the spin-s primary for $s \leqq k/2$ or with the spin $k/2$ primary for $s > k/2$. Thus, the level k Kac-Moody algebra, as in the level k WZW $SU(2)$ model, governs the behaviour of the Kondo state in the presence of k channels of conducting electrons and magnetic impurity of spin s.

This is also the reason why, already in low temperatures, entangled magnetic matter of impurities and conduction electrons indicates the correlation with exotic 4-geometry. First, every CH generates a fermion field. Every small exotic $\mathbb{R}^4$ can be represented as handlebody where Akbulut cork has several CH's attached. The important thing is that the handlenbody has a boundary and only after removing it the interior is diffeomorphic to, say, exotic $\mathbb{R}^4_k$. Let us remove a single CH from the handlebody $\mathbb{R}^4_k$. The result is $\mathbb{R}^4_k \setminus \mathrm{CH}$. The boundary of it reads $\partial(\mathbb{R}^4_k \setminus \mathrm{CH})$. The contribution to the Einstein action $\int_{\mathbb{R}^4_k\setminus\mathrm{CH}} R\sqrt{g}d^4x$ from this boundary is the suitable surface term:

$$\int_{\partial(\mathbb{R}^4_k\setminus\mathrm{CH})} R\sqrt{g}d^4x + \int_{\partial(\mathbb{R}^4_k\setminus\mathrm{CH})} K_{CH}\sqrt{g_\partial}d^3x$$

where K_{CH} is the trace of the 2-nd fundamental form and g_∂ the metric on the boundary Asselmeyer-Maluga & Brans (2011). But as shown in Asselmeyer-Maluga & Brans (2011) this term is expressed by the spinor field ψ describing the immersion of D^2 into $\mathbb{R}^3$, which extends to the immersion of $D^2 \times (0,1)$ into $\mathbb{R}^4$:

$$\int_{\partial(\mathbb{R}^4_k \setminus \mathrm{CH})} K_{CH}\sqrt{g_\partial}d^3x = \int_{\partial(\mathbb{R}^4_k \setminus \mathrm{CH})} \psi\gamma^\mu D_\mu \overline{\psi}\sqrt{g_\partial}d^3x\,. \tag{27}$$

This can be extended to 4-dimensional Einstein-Hilbert action with the source depending on the CH, hence on exotic $\mathbb{R}^4_k$:

$$S_4^{\mathrm{CH}}(\mathbb{R}^4_k) = \int_{\mathbb{R}^4_k \setminus \mathrm{CH}} (R + \psi\gamma^\mu D_\mu \overline{\psi})\sqrt{g}d^4x\,. \tag{28}$$

Again, it was shown in Asselmeyer-Maluga & Brans (2011) that the spinor field ψ extends over whole 4-manifold such that the 4D Dirac equations are fulfilled. This way we have fermion fields which are determined by CH. Moreover, this fermions plays a role of gravity sources as in 28. In fact every infinite branch of the CH determines some 4D fermion.

Second, given exotic $\mathbb{R}^4_p$ we have r Casson handles in its handlebody. These r CH's generate effective $q(r)$-many infinite branches. Each such branch generates a fermion field. Attaching the CH's to the cork results in exotic $\mathbb{R}^4_p$. Hence, p is the function of q in general, $p = p(q(r))$.

Let us assign now the simplest possible CH to every CH in the handlebody of exotic $\mathbb{R}^4$, such that replacing the original CH by this simple one does not change the exotic smoothness. This is the model handlebody we refer to in the context of the Kondo effect (see Fig. 3 for the examples of the simplest CH's).

The k-channel Kondo state, in the k-channel Kondo effect, is the entangled state of conducting electrons in k bands and the magnetic spin s impurity. The physics of resulting state is described by BCFT by the Verlinde fusion rules in $SU(2)_k$ WZW model. To have the WZ term in this WZW model one certainly needs $p = k$. This kWZ term is generated by exotic $\mathbb{R}^4_k$ as we explained in Sec. 2. The draft of the dependance of the number of infinite branches on the function of the number of CH's in the handlebody of $\mathbb{R}^4_k$, is presented in Fig. 3a. Fig. 3b shows the example of the ramified structure of CH's in the precise language of the graphical Kirby calculus (see e.g. Gompf & Stipsicz (1999)).

The general correspondence appears:

One assigns the 4-smooth geometry on $\mathbb{R}^4$ to the k-channel Kondo effect such that k corresponds to the number of infinite branches of CH's in the handlebody. This 4-geometry is $\mathbb{R}^4_p$ where $p = p(k)$, $p, k \in \mathbb{N}$. The change between the physical Kondo states, from this emerging in k_1 channel Kondo effect to this with k_2 channels, $k_1 \neq k_2$, corresponds to the change between 4-geometries, from exotic $\mathbb{R}^4_{p_1}$ to $\mathbb{R}^4_{p_2}$, $p_1 \neq p_2$, $p_1, p_2 \in \mathbb{N}$, such that $p_1 = p_1(k_1)$ and $p_2 = p_2(k_2)$ as above.

Whether actually $p = k$ or not is the question about the level of the $SU(2)$ WZW model and the corresponding fusion rules in use. If $k = p$ the exotic geometry gives the same fusion rules as the Affleck proposed. In the case $k \neq p$ and $k < p$ in the k-channel Kondo effect the fusion rules derived from the exotic geometry are those of the $SU(2)_p$ WZW model. It would be interesting to decide experimentally, which fusion rules apply for bigger k. Probably in higher energies, if the Kondo state survives, the proper fusion rules are those derived from exotic $\mathbb{R}^4_p$. This reflects the situation that electrons in different conduction bands (channels) are generated potentially by (each infinite branch of) Casson handles from the handlebody of

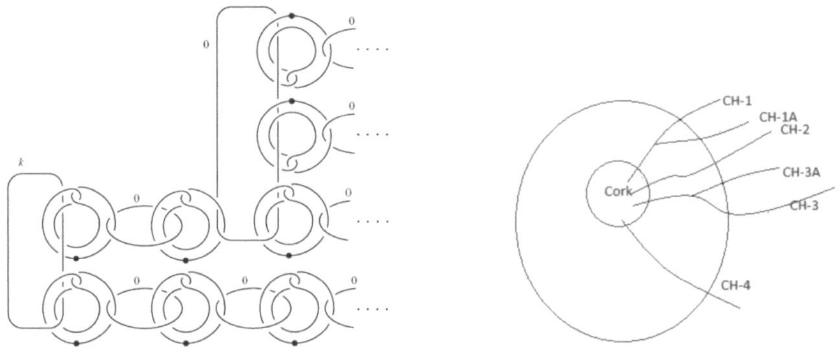

(a) (b)

Fig. 3. (a) (Redrawn from Asselmeyer-Maluga & Brans (2011)) Two CH's, the upper one with 3 infinite branches, the lower one is the simplest CH with the single infinite labeled branch with a single intersection point at every stage. This CH appears in the simplest possible exotic smooth $\mathbb{R}^4$. (b) Schematic structure of the $r = 4$ CH's in the handlebody of exotic $\mathbb{R}^4_p$. Each infinite branch generates a fermion field hence, this exotic $\mathbb{R}^4_p$ can model the Kondo state in the $k = 6$-channels Kondo effect.

the exotic $\mathbb{R}^4_p$, though not every CH generates the actual channel contributing to the Kondo effect. The higher energy more potential CH contributes to the actual electron bands. Then, the fusion rules are given by the exotic geometry, i.e. $SU(2)_p$ WZW model. In suitable high energies 4-geometry (Casson handles) acts as anihilation or creation operator for fermions (electrons). This is the content of our *relativistic fusion rule* (RFR) hypothesis. The experimental confirmation of such discrepancy (between the levels of the WZW model) in high energies in Kondo effect for $p > 2$ channels, would serve as indication for the role of 4-exotic geometry in the relativistic limit of the Kondo state.

Let us illustrate this hypothesis and consider the simplest CH and the simplest exotic $\mathbb{R}^4$ described by Bižaca & Gompf (1996). Suppose this exotic $\mathbb{R}^4_1$ is the member of the radial family and its radius, hence GV invariant of the foliation of S^3, is equal to 1. The corresponding WZ term would be then derived from the $SU(2)_1$ WZW model. Thus, in this case, there is precisely one channel of conducting electrons in the Kondo effect. More complicated exotic $\mathbb{R}^4_2$ could have two CH's in the handlebody and the radii equal to $\sqrt{2}$. Two channels of conducting electrons give rise to the $SU(2)_2$ WZ fusion rules. However, more complicated exotic $\mathbb{R}^4_p$, $p > 2$, could spoil this 1 to 1 correspondence between number of CH's and the number of channels in the Kondo effect.

We have derived the trace of the (exotic) 4-geometry in the low energy Kondo effect. This geometry is probably not physically valid at energies of the Kondo effect (as gravity is not). However, exotic $\mathbb{R}^4_k$ in high energy (and relativistic) limit can become dominating or giving viable physical contributions. These contributions appear when geometric CH's become the actual sources for fermions in KE, thus, changing its CFT structure. In fact the appearance of non-flat $\mathbb{R}^4_k$ when describing the p-channel Kondo effect, indicates a new fundamental link between *matter, geometry* and *gravity* in dimension 4.

5. From smooth geometry of string backgrounds to quantum D-branes

One could wonder what is, if any, suitable sense assigned to geometry of spacetime in various string constructions or backgrounds. As we know the geometry of GR, hence, classical gravity, is the one of (pseudo)-Riemannian differentiable manifolds. String theory has GR (10D Einstein equations) as its classical gravitational limit; however, string theory is the theory of QG and the spacetime geometry should be modified. What is the fate of this (pseudo) Riemannian geometry when gravity is quantized? To answer this question we should find correct classical limit for some quantum string constructions. The proper way is to consider the string backgrounds. These are semi-classical solutions in string theory or supergravity, around which one develops a purturbative theory. GR is not the only ingredient of classical geometry in string theory. There are other fields which are equally fundamental. In type II we have metric $G_{\mu\nu}$, antisymmetric H-field, i.e. three-form $H_{\mu\nu\rho}$, and dilaton Φ. In heterotic strings we have additionally gauge field $F^a_{\mu\nu}$ and the calculations of Sec. 3 made use of these. The presence of B-field such that H is represented by the non-zero cohomology class (see below), is a highly non-trivial fact and indicates that the correct, semi-classical, geometry for string theory is one based on *abelian gerbes* as supplementing Riemannian geometry Król (2010a;b); Segal (2001). Small exotic $\mathbb{R}^4$'s show strong connections with abelian gerbes on S^3 Asselmeyer-Maluga & Król (2009a) which has many important consequences. Some of them are discussed in what follows.

Another crucial feature is the role assigned to D- and NS-branes. Closed string theory, as we made use of it in Sec. 3, is not complete in the sense that there are possible boundary conditions, Dirichlet (D) or Neveu-Schwarz (NS), for open strings, already appearing in closed string theories. These boundary conditions determine geometric subspaces on which open strings can end. In that sense open string theory complements the closed one and predicts the existence of D- or NS-branes. This tame picture of branes as subspaces has only very limited validity. In the quantum regime, or even in the non-zero string coupling g_s, the picture of D-branes as above fails Aspinwall (2004). Nevertheless, interesting proposals were presented recently. They are based on the ideas from non-commutative geometry and aim toward replacing D-branes and spacetime by corresponding (sub) $C^\star$-algebras Brodzki et al. (2008a;b); Szabo (2008). Surprisingly, such an $C^\star$-algebraic setting again shows deep connections with exotic $\mathbb{R}^4$'s.

The appearance of the codimension-one foliations of S^3 in the structure of small exotic $\mathbb{R}^4$, is the key for the whole spectrum of the connections of exotics, beginning with differential geometry and topology, up to non-commutative geometry. This opens very atractive possibilities for exploring both, 1) the classical limit of string geometry, as above and 2) quantum D-branes regim in string theory.

Let us comment on 1) above. The presence of non-zero B-field in a string background is crucial from the point of view of resulting geometry: in σ-model the B-field modifies metric as in (3). Moreover, supposing dilaton is constant and $F^a_{\mu\nu}$ vanishes, the second equation of (8) (the β-function), enforces the background be non-flat, unless $H = dB$ is zero. Given S^3 part of the linear dilaton background as in Sec. 3, we have non-trivial H-field on it. The topological classification of H-fields is given by 3-rd de Rham cohomology classes on background manifold M, $H^3(M, \mathbb{R})$. In order to avoid anomalies we restrict to the integral case $H^3(S^3, \mathbb{Z})$ for $M = S^3$. These classes however are equally generated by exotic $\mathbb{R}^4_k, k \in \mathbb{Z}$ (see Sec. 2). On the other hand, the classification of D-branes in string backgrounds is governed by K-theory of the background, or in the presence of H-field, by, twisted by H, K-theory classes. This is briefly summerized in the next subsection where D and NS branes will be understood also classically as subsets in specific CFT backgrounds.

5.1 NS and D branes in type II

Let us consider again the bosonic, i.e. nonsupersymmetric, $SU(2)_k$ WZW model and follow Asselmeyer-Maluga & Król (2011c) closely. The semi-classical limit of it corresponds to taking $k \to \infty$ as in Sec. 3.3. In that limit D-branes in group manifold $SU(2)$ are determined by wrapping the conjugacy classes of $SU(2)$, i.e. are described by 2-spheres S^2's and two poles (degenarate branes) each localized at a point. Owing to the quantization conditions, there are $k+1$ D-branes on the level k $SU(2)$ WZW model Alekseev & Schomerus (1999b); Fredenhagen & Schomerus (2001); Schomerus (2002). To grasp the dynamics of the branes one should deal with the gauge theory on the stack of N D-branes on S^3, quite similar to the flat space case where noncommutative gauge theory emerges Alekseev & Schomerus (1999a). Let J be the representation of $SU(2)_k$ i.e. $J = 0, \frac{1}{2}, 1, ..., \frac{k}{2}$. The non-commutative action for the dynamics of N branes of type J (on top of each other), in the string regim (k is finite), is then given by:

$$S_{N,J} = S_{YM} + S_{CS} = \frac{\pi^2}{k^2(2J+1)N}\left(\frac{1}{4}\mathrm{tr}(F_{\mu\nu}F^{\mu\nu}) - \frac{i}{2}\mathrm{tr}(f^{\mu\nu\rho}\mathrm{CS}_{\mu\nu\rho})\right). \tag{29}$$

Here the curvature form $F_{\mu\nu}(A) = iL_\mu A_\nu - iL_\nu A_\mu + i[A_\mu, A_\nu] + f_{\mu\nu\rho}A^\rho$ and the noncommutative Chern-Simons action reads $\mathrm{CS}_{\mu\nu\rho}(A) = L_\mu A_\nu A_\rho + \frac{1}{3}A_\mu[A_\nu, A_\rho]$. The fields A_μ, $\mu = 1,2,3$ are defined on a fuzzy 2-sphere S^2_J and should be considered as $N \times N$ matrix-valued, i.e. $A_\mu = \sum_{j,a} \mathrm{a}^\mu_{j,a} Y^j_a$ where Y^j_a are fuzzy spherical harmonics and $\mathrm{a}^\mu_{j,a}$ are Chan-Paton matrix-valued coefficients. L_μ are generators of the rotations on fuzzy 2-spheres and they act only on fuzzy spherical harmonics Schomerus (2002). The noncommutative action S_{YM} was derived from Connes spectral triples from the noncommutative geometry, and they will be crucial in grasping quantum nature of D-branes in the next subsection. Originally the action (29) was designed to describe Maxwell theory on fuzzy spheres Carow-Watamura & Watamura (2000). The equations of motion derived from (29) read:

$$L_\mu F^{\mu\nu} + [A_\mu, F^{\mu\nu}] = 0\,. \tag{30}$$

The solutions of (30) describe the dynamics of the branes, i.e. the condensation processes on the brane configuration (N, J) which results in another configuration (N', J'). A special class of solutions, in the semi-classical $k \to \infty$ limit, can be obtained from the $N(2J+1)$ dimensional representations of the algebra su(2). For $J = 0$ one has N branes of type $J = 0$, i.e. N point-like branes in S^3 at the identity of the group. Given another solution corresponding to $J_N = \frac{N-1}{2}$, one shows that this solution is the condensed state of N point-like branes at the identity of $SU(2)$ Schomerus (2002):

$$(N, J) = (N, 0) \to (1, \frac{N-1}{2}) = (N', J') \tag{31}$$

Turning to the finite k string regime of the $SU(2)$ WZW model one makes use of the techniques of the boundary CFT, the same as was applied to the analysis of Kondo effect in Sec. 4. It follows that there exists a continuous shift between the partition functions governed by the Verlinde fusion rules coefficients $N^l_{J_N j}$: $N\chi_j(q)$ and the sum of characters $\sum_j N^l_{J_N j}\chi_l(q)$ where

$N = 2J_N + 1$. In the case of N point-like branes one can determine the decay product of these by considering open strings ending on the branes. The result on the partition function is

$$Z_{(N,0)}(q) = N^2\chi_0(q)$$

which is continuously shifted to $N\chi_{J_N}(q)$ and next to $\sum_j N^{\,j}_{J_N J_N}\chi_j(q)$. As the result, we have the decay process:

$$\begin{aligned} Z_{(N,0)}(q) &\to Z_{(1,J_N)} \\ (N,0) &\to (1,J_N) \end{aligned} \tag{32}$$

which extends the similar process derived at the semi-classical $k \to \infty$ limit (31), and the representations $2J_N$ are bounded now, from the above, by k.

Thus, there are $k+1$ stable branes wrapping the conjugacy classes numbered by $J = 0, \frac{1}{2}, ..., \frac{k}{2}$. The decaying process (32) says that placing N point-like branes (each charged by the unit 1) at the pole e, they can decay to the spherical brane J_N wrapping the conjugacy class. Taking more point-like branes to the stack at e, gives the more distant S^2 branes until reaching the opposite pole $-e$, where we have single point-like brane with the opposite charge -1. Having identified $k+1$ units of the charge with -1, we obtain the correct shape of the group of charges, as: $\mathbb{Z}_{k+2}$. More generally, the charges of branes on the background X with non-vanishing $H \in H^3(X,\mathbb{Z})$ are described by the twisted K group, $K^{\star}_H(X)$. In the case of $SU(2)$, we get the group of RR charges as (for $K = k+2$):

$$K^{\star}_H(S^3) = \mathbb{Z}_K \tag{33}$$

Now, based on the earlier discussion from Secs. 2,3, let us place the $S^3 \simeq SU(2)$ above, at the boundary of the Akbulut cork for some exotic smooth $\mathbb{R}^4_k$. Then, we have: *Certain small exotic $\mathbb{R}^4$'s generate the group of RR charges of D-branes in the curved background of $S^3 \subset \mathbb{R}^4$.*

We have yet another important correspondence:

Theorem 2 (Asselmeyer-Maluga & Król (2011c)). *The classification of RR charges of the branes on the background given by the group manifold $SU(2)$ at the level k (hence the dynamics of D-branes in S^3 in stringy regime) is correlated with the exotic smoothness on $\mathbb{R}^4$, containing this $S^3 = SU(2)$ as the part of the boundary of the Akbulut cork.*

Turning to the linear dilaton geometry, as emerging, in the near horizon geometry, from the stack of N NS5-branes in supersymmetric model, i.e. $\mathbb{R}^{5,1} \times \mathbb{R}_\phi \times SU(2)_k$, we obtain next important relation:

Theorem 3 (Asselmeyer-Maluga & Król (2011c)). *In the geometry of the stack of NS5-branes in type II superstring theories, adding or subtracting a NS5-brane is correlated with the change of the smoothness structure on the transversal $\mathbb{R}^4$.*

5.2 Quantum and topological D-branes

The recognition of the role of exotic $\mathbb{R}^4$ in string theory, in the previous and in Sec. 3, relied on the following items:

- Standard smooth $\mathbb{R}^4$ appears as a part of an exact string background;

- The process of changing the exotic smoothness on $\mathbb{R}^4$ is capable of encoding a) the change in the configurations of specific D- and NS branes (Sec. 5.1), b) the change of the 4D part of the string background from flat to curved one in closed string theory (see Sec. 3).
- All exotic $\mathbb{R}^4$'s appearing in this setup are *small exotic* $\mathbb{R}^4$'s, i.e. those which embed smoothly in the standard smooth $\mathbb{R}^4$ as open subsets.

Given the fact that every small exotic $\mathbb{R}^4$ from the radial family (see Sec. 2) determines the codimension-1 foliation of S^3, we have natural $C^\star$-algebra assigned to this 4-exotic. Namely this is the noncommutative convolution $C^\star$-algebra of the foliation. Let us, following Brodzki et al. (2008b), represent every D-brane by suitable separable $C^\star$-algebra replacing, in the same time, spacetime by the correspponding separable $C^\star$-algebra as well. The usual semiclassical embedding of D-branes in spacetime is now reformulated in the language of morphisms between $C^\star$-algebras. In fact, taking into account the isue of stability of D-branes, we define the setup:

1. Fix the (spacetime) $C^\star$ algebra $\mathcal{A}$;
2. A $\star$ homomorphism $\phi : \mathcal{A} \to \mathbf{B}(\mathcal{H})$ (a homomorphisms of the algebras preserving their $\star$ structure), generates the embedding of the D-brane world-volume M and its noncommutative algebra $\mathcal{A}_M$ as $\mathcal{A}_M := \phi(\mathcal{A})$;
3. D-branes embedded in a spacetime $\mathcal{A}$ are represented by the spectral triple $(\mathcal{H}, \mathcal{A}_M, T)$;
4. Equivalently, a D-brane in A is given by an unbounded Fredholm module $(\mathcal{H}, \phi, T)$.

Thus, the classification of stable D-branes in $\mathcal{A}$ is given by the classification of Fredholm modules $(\mathcal{H}, \phi, T)$ where $\mathbf{B}(\mathcal{H})$ are bounded operators on the separable Hilbert space $\mathcal{H}$ and T the operators corresponding to tachyons. In general, to every foliation (V, F) one can associate its noncommutative $C^\star$ convolution algebra $C^\star(V, F)$. The interesting connection with exotic 4-smoothness then emerges:

Theorem 4. *The class of generalized stable D-branes on the $C^\star$ algebra $C^\star(S^3, F_1)$ (of the codimension 1 foliation of S^3) determines an invariant of exotic smooth $\mathbb{R}^4$,*

and

Theorem 5. *Let e be an exotic $\mathbb{R}^4$ corresponding to the codimension-1 foliation of S^3 which gives rise to the $C^\star$algebra $\mathcal{A}_e$. The exotic smooth $\mathbb{R}^4$ embedded in e determines a generalized quantum D-brane in $\mathcal{A}_e$.*

It is interesting to note that the tame subspace interpretation of D-branes can be recovered for the special class of the topological quantum D-branes. However, the embedding is replaced now by the wild embedding into spacetime, which historically appeared in the description of the horned Alexander's spheres, known from topology.

Theorem 6. *Let $\mathbb{R}^4_H$ be some exotic $\mathbb{R}^4$ determined by element in $H^3(S^3, \mathbb{R})$, i.e. by a codimension-1 foliation of S^3. Each wild embedding $i : K^3 \to S^p$ for $p > 6$ of a 3-dimensional polyhedron determines a class in $H^n(S^n, \mathbb{R})$ which represents a wild embedding $i : K^p \to S^n$ of a p -polyhedron into S^n.*

Now, a class of *topological quantum Dp-branes* are these branes which are determined by the wild embeddings $i : K^p \to S^n$ as above and in the classical and flat limit correspond to tame embeddings. In fact, B-field on S^3 can be translated into wild embeddings of higher dimensional objects and generates quantum character of these branes.

6. Discussion and conclusions

Superstring theory (ST) appears in fact as very rich mathematics. The mathematics which is designed especially for the reconciling classical gravity, as in GR, with QFT. The richness of mathematics involved is, however, the limitation of the theory. Namely, to yield 4D physics from such huge structure is very non-unique and thus problematic. We followed the idea, proposed at the recent International Congress of Mathematician ICM 2010 [Asselmeyer-Maluga & Król (2010)], that the mathematics of ST refers to and advance understanding of the mathematics of exotic smooth $\mathbb{R}^4$. Conversely, exotic $\mathbb{R}^4$'s provide important information about the mathematics of superstrings. Exotic $\mathbb{R}^4$'s are non-flat geometries, hence contain gravity from the point of view of physics. ST is the theory of QG and gravity of exotic geometries is quantized by methods of ST. The 4-geometries also refer to effective correlated states of condensed matter as in Kondo effect. Thus, the approach presented in this Chapter indicates new fundamental link between *gravity, geometry and matter* at the quantum limit and exclusively in dimension 4. The exotic smoothness of $\mathbb{R}^4$, when underlies the 4-Minkowski spacetime, is a natural way to quantum gravity (given by superstring techniques) from the standard model of particles. On the other hand, exotic $\mathbb{R}^4$'s serve as factor reducing the ambiguity of 10D superstring theory in yielding 4D physical results. The work on these issues should be further pursued.

7. References

Affleck, I. (1995). Conformal field theory approach to the Kondo effect, *Acta Phys. Pol. B* 26 6: 1869.

Affleck, I. & Ludwig, A. (1991). Exact, asymptotic, three-dimensional, space- and time-dependent, Green's functions in the multichannel Kondo effect, *Phys. Rev, Lett.* 67: 3160–3163.

Affleck, I. & Ludwig, A. (1993). Exact conformal field theory results on the multichannel Kondo effect: single-fermion Green function, self-energy, and resistivity, *Phys. Rev. B* 48: 7297–7321.

Affleck, I. & Ludwig, A. (1994). Exact conformal-field-theory results on the multi-channel Kondo effect: Asymptotic three-dimensional space- and time-dependent multi-point and many-particle Greens functions, *Nucl. Phys. B* 428: 545–611.

Akbulut, S. (1996). Lectures on Seiberg-Witten invariants, *Turkish J. Math.* 20: 95–119.

Akbulut, S. & Yasui, K. (2008). Corks, plugs and exotic structures, *Journal of Gokova Geometry Topology* 2: 40–82. arXiv:0806.3010.

Akbulut, S. & Yasui, K. (2009). Knotted corks, *J Topology* 2: 823–839. arXiv:0812.5098.

Alekseev, A., R. V. & Schomerus, V. (1999a). Non-commutative world-volume geometries: Branes on SU(2) and fuzzy spheres, *JHEP* 9909: 023.

Alekseev, A. & Schomerus, V. (1999b). D-branes in the WZW model, *Phys. Rev. D* 60: 061901.

Anderson, P. (1961). Localized magnetic states in metals, *Phys. Rev.* 124: 41–53.

Aspinwall, P. S. (2004). D-Branes on Calabi-Yau manifolds. arXiv:hep-th/0403166.

Asselmeyer-Maluga, T. (2010). Exotic smoothness and quantum gravity, *Class. Q. Grav.* 27: 165002. arXiv:1003.5506.

Asselmeyer-Maluga, T. & Brans, C. (2007). *Exotic Smoothness and Physics*, WorldScientific Publ., Singapore.

Asselmeyer-Maluga, T. & Brans, C. H. (2011). Gravitational sources induced by exotic smoothness. arXiv:1101.3168.

Asselmeyer-Maluga, T., G. P. & Król, J. (2011a). The modification of the energy spectrum of charged particles by exotic open 4-smoothness via superstring theory. arXiv: 1109.1973.

Asselmeyer-Maluga, T. & Król, J. (2009a). Abelian gerbes, generalized geometries and exotic R^4. arXiv: 0904.1276.

Asselmeyer-Maluga, T. & Król, J. (2009b). Gerbes on orbifolds and exotic smooth R^4. arXiv: 0911.0271.

Asselmeyer-Maluga, T. & Król, J. (2010). Small exotic smooth R^4 and string theory, *International Congress of Mathematicians ICM 2010 Short Communications Abstracts Book*, Ed. R. Bathia, Hindustan Book Agency, p. 400.

Asselmeyer-Maluga, T. & Król, J. (2011b). Constructing a quantum field theory from spacetime. arXiv:1107.3458.

Asselmeyer-Maluga, T. & Król, J. (2011c). Exotic smooth R^4 and certain configurations of NS and D branes in string theory, *Int. J. Mod. Phys. A* 26: 1375–1388.

Asselmeyer-Maluga, T. & Król, J. (2011d). Topological quantum D-branes and wild embeddings from exotic smooth R^4, *Int. J. Mod. Phys. A* 26: 3421–3437. arXiv:1105.1557.

Asselmeyer-Maluga, T. & Król, J. (2012). Quantum D-branes and exotic smooth R^4, will appear in, *Int. J. Geom. Meth. Mod. Phys.* 9. arXiv:1102.3274.

Asselmeyer, T. (1996). Generation of source terms in general relativity by differential structures, *Class. Quant. Grav.* 14: 749 – 758.

Bižaca, Ž. (1994). A reimbedding algorithm for Casson handles, *Trans. Amer. Math. Soc.* 345: 435–510.

Bižaca, Ž. & Gompf, R. (1996). Elliptic surfaces and some simple exotic $\mathbb{R}^4$'s, *J. Diff. Geom.* 43: 458–504.

Brans, C. (1994a). Exotic smoothness and physics, *J. Math. Phys.* 35: 5494–5506.

Brans, C. (1994b). Localized exotic smoothness, *Class. Quant. Grav.* 11: 1785–1792.

Brodzki, J., Mathai, V., Rosenberg, J. & Szabo, R. J. (2008a). D-branes, KK-theory and duality on noncommutative spaces, *J. Phys. Conf. Ser.* 103: 012004. arXiv:hep-th/0709.2128.

Brodzki, J., Mathai, V., Rosenberg, J. & Szabo, R. J. (2008b). D-branes, RR-fields and duality on noncommutative manifolds, *Commun. Math. Phys.* 277: 643. arXiv:hep-th/0607020.

Carow-Watamura, J. & Watamura, S. (2000). Noncommutative geometry and gauge theory on fuzzy sphere, *Commun. Math. Phys.* 212: 395. arXiv:hep-th/9801195.

Casson, A. (1986). *Three lectures on new infinite constructions in 4-dimensional manifolds*, Vol. 62, Progress in Mathematics edn, Birkhäuser. Notes by Lucian Guillou, first published 1973.

Curtis, C., F. M. H. W. & Stong, R. (1997). A decomposition theorem for h-cobordant smooth simply connected compact 4-manifolds, *Inv. Math.* 123: 343–348.

DeMichelis, S. & Freedman, M. (1992). Uncountable many exotic $\mathbb{R}^4$'s in standard 4-space, *J. Diff. Geom.* 35: 219–254.

Donaldson, S. & Kronheimer, P. (1990). *The Geometry of Four-Manifolds*, Oxford Univ. Press, Oxford.

Fredenhagen, S. & Schomerus, V. (2001). Branes on group manifolds, gluon condensates, and twisted K-theory, *JHEP* 04: 007. arXiv:hep-th/0012164.

Freedman, M. (1979). A fake $S^3 \times R$, *Ann. of Math.* 110: 177–201.

Freedman, M. (1982). The topology of four-dimensional manifolds, *J. Diff. Geom.* 17: 357 – 454.

Gompf, R. (1984). Infinite families of Casson handles and topological disks, *Topology* 23: 395–400.

Gompf, R. (1989). Periodic ends and knot concordance, *Top. Appl.* 32: 141–148.

Gompf, R. & Stipsicz, A. (1999). *4-manifolds and Kirby Calculus*, American Mathematical Society.

Hassan, S. & Sen, A. (1993). Marginal deformations of WZNW and coset models from O(d, d) transformation, *Nucl. Phys. B* 405: 143. arXiv:hep-th/9210121.

Kiritsis, E. & Kounnas, C. (1995a). Curved four-dimensional spacetime as infrared regulator in superstring theories, *Nucl. Phys.Proc.Suppl. 41 (1995) 331-340* 41: 331–340. arXiv:hep-th/9410212.

Kiritsis, E. & Kounnas, C. (1995b). Infrared behavior of closed superstrings in strong magnetic and gravitational fields, *Nucl. Phys. B* 456: 699–731. arXiv:hep-th/9508078.

Kiritsis, E. & Kounnas, C. (1995c). Infrared-regulated string theory and loop corrections to coupling constants. CERN-TH/95-172, LPTENS-95/29, arXiv:hep-th/9507051.

Król, J. (2004a). Background independence in quantum gravity and forcing constructions, *Found. Phys.* 34: 361–403.

Król, J. (2004b). Exotic smoothness and non-commutative spaces. The model-theoretic approach, *Found. Phys.* 34: 843–869.

Król, J. (2005). Model theory and the AdS/CFT correspondence. presented at the IPM String School and Workshop, Queshm Island, Iran, 05-14. 01. 2005, arXiv:hep-th/0506003.

Król, J. (2010a). Exotic smooth 4-manifolds and gerbes as geometry for quantum gravity, *Acta. Phys. Pol. B* 40 11.

Król, J. (2010b). (Quantum) gravity effects via exotic $\mathbb{R}^4$, *Ann. Phys. (Berlin)* 19: No. 3–5, 355–358.

Król, J. (2011a). Quantum gravity and quantum matter in 4-dimensions, *Acta. Phys. Pol. B* 42 11: 2335-2342.

Król, J. (2011b). New 4D results from superstring theory, *Acta. Phys. Pol. B* 42 11: 2343-2350.

Orlando, D. (2006). Corfu 05 lectures - part I: Strings on curved backgrounds, *J.Phys.Conf.Ser.* 53: 551. arXiv:hep-th/0607027.

Orlando, D. (2007). *String Theory: exact solutions, marginal deformations and hyperbolic spaces*, PhD thesis, Universty Aarhus. Fortsch.Phys. 55:161-282, arXiv:hep-th/0610284.

Potok, R., Rau, I., Shtrikman, H., Oreg, Y. & Goldhaber-Gordon, D. (2007). Observation of the two-channel Kondo effect, *Nature* 446: 167–171.

Prezas, N. & Sfetsos, K. (2008). Supersymmetric moduli of the $SU(2) \times R^{\phi}$ linear dilaton background and NS5-branes, *JHEP06(2008)080* . http://iopscience.iop.org/1126-6708/2008/06/080.

Schomerus, V. (2002). Lectures on branes in curved backgrounds. arXiv:hep-th/0209241.

Scorpan, A. (2005). *The wild world of 4-manifolds*, AMS, USA.

Segal, G. B. (2001). Topological structures in string theory, *Phil. Trans. Roy. Soc. London* A 359: 1389.

Sładkowski, J. (2001). Gravity on exotic $\mathbb{R}^4$ with few symmetries, *Int.J. Mod. Phys. D* 10: 311–313.

Szabo, R. J. (2008). D-branes and bivariant K-theory. Based on invited lectures given at the workshop "Noncommutative Geometry and Physics 2008 - K-Theory and D-Brane-",

February 18-22 2008, Shonan Village Center, Kanagawa, Japan. To be published in the volume Noncommutative Geometry and Physics III by World Scientific, arXiv: 0809.3029.
Tamura, I. (1992). *Topology of Foliations: An Introduction*, Vol. 97 of *Translations of Math. Monographs*, AMS, Providence.
Thurston, W. (1972). Noncobordant foliations of S^3, *BAMS* 78: 511 – 514.
Witten, E. (1985). Global gravitational anomalies, *Commun. Math. Phys.* 100: 197–229.

5

Fiber Bundles, Gauge Theories and Gravity

Rodrigo F. Sobreiro
UFF - Universidade Federal Fluminense, Instituto de Física, Campus da Praia Vermelha, Niterói, Brasil

1. Introduction

Motivated by the construction of a gravity theory independently of the metric structure of spacetime and on the stability of a quantum gravity theory many authors have developed schemes that allow a gauge theory to generate an effective metric, see for instance [1–9]. The models are constructed based on a gauge group G that possesses the Lorentz group $SO(1,3)$ as a stable subgroup. A symmetry breaking mechanism is imposed in order to G collapse to $SO(1,3)$. Mostly of these techniques are based on the de Sitter group and its variations. However, other groups are also considered such as the general linear and affine groups, see for instance [10–14], and also unitary groups [15]. The main motivation in the construction of a gauge theory of gravity that is metric independent is that the base space can be regarded as a flat one, and thus the standard quantization of gauge theories can be employed [16]. In fact, some of the cited works are in fact quantizable, at least perturbatively.

In the present work we consider the fiber bundle theory to describe gauge theories and gravity [17–20]. We then show that a gauge theory can be identified with a first order gravity if the principal bundle that describes the gauge theory can be identified with the principal bundle that describes gravity. We formally establish the conditions that the gauge theory must obey and the resulting gravity theory that emerges. The last is constructed from a mapping between the gauge principal bundle structures and the geometric setting of a gravity theory.

This work is organized as follows: In Sect. 2 we briefly review the fiber bundle description of gauge theories. Also in this section we enunciate some important results concerning reduction of principal bundles. The same approach to the first order gravity theories is displayed in Sect. 3. In Sect. 4 we discuss the emergent geometries that can be derived from a gauge theory in terms of formal theorems. In sect. 5 we collect our final remarks.

2. Gauge theories

2.1 Principal bundles for gauge theories

First we define two classes of principal bundles within gauge theories can be formally described. The first one is the principal bundle which localizes a gauge group [18] $G_R = (G, R)$ where G is a Lie group characterizing the fiber and structure group while R is the base space, a differential manifold with d_o dimensions identified with spacetime. The total space G_R describes the localization of the Lie group G in the manifold R, assembling to each point $x \in R$ a different value for the elements of G. We shall refer to G_R as *gauge bundle*.

It is assumed that G_R is endowed with a connection 1-form Y. The connection 1-form is recognized as the gauge field, the fundamental field of gauge theories. The connection

1-form will be called gauge connection, or simply, connection. The gauge transformations are associated with coordinates changing of the total space with fixed base space coordinates, $(x,g) \to (x,g')$, which corresponds to a translation along the fiber, providing $Y(x,g) \longmapsto Y(x,g') = f^{-1}(x)\,(\mathrm{d} + Y(x,g))\,f(x)$, where $g' = gf$ and $\{g',g,f\} \subset G$. To every connection Y there is a curvature 2-form defined over G_R, namely $F = \nabla^2 = \mathrm{d}Y + YY$, where ∇ is the covariant derivative, $\nabla = \mathrm{d} + Y$, and d is the exterior derivative in R. The covariant derivative is defined from the parallel transport between fibers and the curvature is obviously recognized as the field strength in gauge theories.

The gauge connection does not belong to the former structure of the gauge bundle, it originates from a unique choice for the decomposition of the tangent space $T_q(G_R)$, in a point $q \in G_R$, into vertical and horizontal spaces. The mathematical structure that describes the dynamics of Y must contain all possible gauge connections that can be defined in G_R as well as the information of gauge transformations as the definition of equivalence classes for gauge connections. This task is achieved through the moduli bundle $\mathbb{Y} = (G_R, \mathcal{Y})$, see for instance [14, 18, 20–22]. In $\mathbb{Y}$, the fiber and structure group are the local Lie group G_R and the base space $\mathcal{Y}$ is the space of all independent connection 1-forms[1] Y, the so called moduli space. The typical fiber[2] $\pi^{-1}(Y)$ is a gauge orbit obtained from a configuration $Y(x) \in \mathcal{Y}$ and all of its possible gauge transformations $Y^g = g^{-1}(\mathrm{d} + Y)g$. Thus, the total space $\mathbb{Y}$ can be understood as the union of all gauge orbits which determine the equivalence classes in $\mathbb{Y}$.

The interpretation of the gauge and moduli principal bundles is as follows: The gauge bundle provides the localization of a Lie group and the existence of a gauge connection. To give dynamics for the connection one should consider all possible connections (together with a minimizing principle for a classical theory or a path integral measure for a quantum one [14, 22]). This dynamics is provided by the infinite dimensional moduli bundle.

2.2 Contraction of principal bundles

We now discuss some relevant results concerning gauge bundles:

Theorem 2.1. *Let $H_R = (H,R)$ be a reduced gauge bundle obtained from a former gauge bundle $G_R = (G,R)$, where $G = H \otimes K$ induces a Lie algebra decomposition $\widetilde{G} = \widetilde{H} \oplus \widetilde{K}$. If G_R is endowed with a connection form $Y = A + B$, where $A \in \widetilde{H}$ and $B \in \widetilde{K}$, then A defines a connection on H_R if, and only if, H is a stability group of G.*

Comment. This theorem[3] is a standard result [17, 23]. The formal proof can be found in [17]. It follows from the fact that a gauge transformation on a fiber $\pi^{-1}(x)$ will always keep A as a connection and B as an element of $\widetilde{K}$ as it can be seen from the decomposition of the gauge transformation in G_R. Obviously, this is a direct consequence of the stability of H. This result establishes that the original bundle imposes a connection on the reduced bundle, independently of the mechanism that led to H_R.

Corollary 2.2. *The space K defines an associated bundle $K_R = (H,R,K) \equiv H_R \times K$.*

[1] By independent we mean the set of gauge connections that cannot be related to each other through a gauge transformation, *i.e.*, they do not belong to the same equivalence class.
[2] We adopt the standard fiber bundle notation where $\pi : \mathbb{Y} \longmapsto \mathcal{Y}$ is the projection map.
[3] From now on the conditions for the validity of this theorem are assumed to hold.

Proof. The coset K is an invariant subspace with respect to the stability group H and thus a homogeneous space, which is the requirement for K to be the fiber of an associated bundle [20]. From Theorem 2.1 it is clear that a point $q \in G_R$ will split $q = (u, k)$ where $u \in H_R$ and $k \in \widetilde{K}$. Thus, we define the action of H on $H_R \times K$ by $(u, k) \longmapsto (uh, h^{-1}k)$ by taking the transitions functions to act on the fiber K while an element of the group suffers its own action from the right as allowed by the principal bundle nature of H_R. Ever since the point $x \in R$ is general, the proof holds for the entire bundle K_R. □

Comment. From Corollary 2.2 it is clear that the field B is a section over K_R [17, 18]. Thus, the component $B \in \widetilde{K}$ of the connection Y migrates to the sector of matter fields on H_R.

We consider now moduli bundles:

Theorem 2.3. *Let $\mathbb{Y} = (G_R, \mathcal{Y})$ be a moduli bundle constructed from $G_R = (G, R)$. Then the reduction $G_R \longrightarrow H_R$ induces a reduction on $\mathbb{Y}$ according to $\mathbb{Y} \longrightarrow \mathbb{A}$ where the reduced moduli bundle is $\mathbb{A} = (H_R, \mathcal{C})$. The base space $\mathcal{C} = \mathcal{A} \times \mathcal{B}$ is the decomposed moduli space of stable connections $A \in \widetilde{H}$ and independent sections $B \in \widetilde{K}$ on K_R while the fiber is the decomposed gauge orbit:*

$$
\begin{aligned}
A^h &= h^{-1}(\mathrm{d} + A)h\,, \\
B^h &= h^{-1}Bh\,.
\end{aligned} \tag{1}
$$

Proof. Since G_R is the fiber of $\mathbb{Y}$ its reduction to H_R is equivalent to a split on the gauge orbit (1). Thus, the gauge orbit is reduced to the first of (1) where, $\mathcal{A} \subset \mathcal{Y}$, represented by independent elements $A \in \widetilde{H}$, define the reduced moduli space of connections. The space $\mathcal{B} = \mathcal{Y}/\mathcal{A}$, on the other hand, is the set of all fields B that cannot be related through a gauge transformation. Thus, a point in the base space can be defined as $C = (A, B)$ and the fiber is constructed by the action of H as $C \longmapsto C^h = (A^h, B^h)$. The reduced total space is the union of all reduced gauge orbits. The stable character of H ensures that there will be no mixing between the spaces $\mathbb{A}$ and $\mathbb{B} = \mathbb{Y}/\mathbb{A}$ along any gauge orbit. □

Comment. The infinite dimensional space $\mathcal{B}$ is equivalent to the set of all independent sections $B(x)$ that can be defined in K_R. Thus, the space $\mathbb{B}$ is the collection of all possible sections in K_R. The space $\mathbb{B}$ can be also understood as the fiber bundle $\mathbb{B} = (\Sigma(B), H_R, \mathcal{B})$ where the base space is $\mathcal{B}$ and a fiber $\Sigma(B)$ is the collection of all equivalent sections for a given $B \in \mathcal{B}$.

Corollary 2.4. *Define a composite field θ, which is an invariant representation of H, that can be constructed from the original set of connections. For each base space point C there is only one field $\theta(C)$. If an equivalence class C^h is defined then $\theta^h = \theta(C^h)$ is on the same equivalence class of $\theta(C)$ where $\theta^h = h\theta$.*

Proof. The field θ is, by construction, an invariant representation of H, thus, it transforms as $\theta \longmapsto \theta^h = h\theta$. The last expression defines the equivalence class for θ. Now, since $\theta = \theta(C)$, then $\theta(C')$, constructed in another point C', belongs to the same equivalence class of the original field if $\theta(C') = h\theta(C)$. However, the transformation of θ is induced by the action of the group on its dependence on C. Thus, $\theta(C') = \theta(C^g)$ for an element $g \in H$. Using again the definition of θ as an invariant representation, we have that $\theta(C^g) = g\theta(C)$. Thus, $g = h$. □

Comment. The field θ is a one to one map $\theta : C \longmapsto \theta(C)$ which establishes that for at each point C there will be only one θ such that if $C \sim C'$ then $\theta \sim \theta'$. In other words, in each fiber C^h there will be only one equivalence class for θ.

3. First order gravity

Gravity can be mathematically defined as a coframe bundle [12, 13, 20, 21], $C_M = (GL(d,\mathbb{R}), M)$, where M is a d-dimensional spacetime manifold. The structure group and fiber have a more deep meaning: In each point $X \in M$ one can define the cotangent space $T^*_X(M)$. The fiber is the collection of all coframes e that can be defined in $T^*_X(M)$ and which are related to each other through the action of the general linear group. As a consequence, the fiber is actually the group $GL(d,\mathbb{R})$. In terms of Sect. 2, the coframe bundle is also a gauge bundle for the general linear group with the addend that the gauge group is identified with geometric properties of M. The action of the group from the right are local gauge transformations while the action of the group from the left are general coordinate transformations.

Geometrically, the gauge connection Γ is related to the parallel transport, in M, between two near cotangent spaces. The curvature 2-form is obtained from the double action of the covariant derivative, $\Omega = d\Gamma + \Gamma\Gamma$ while torsion, $T = \nabla e$, is the minimal coupling of coframes. It is evident that, besides Γ, which is the gauge field of gravity, e is just as relevant. Moreover, the metric tensor m in the tangent space $T(M)$ has to be introduced because the $GL(d,\mathbb{R})/SO(d-n,n)$ sector of the general linear group does not preserve a flat metric. In practice m enters as an extra independent field. Thus, in C_M, gravity possesses three fundamental fields, Γ, e and m, all relevant to determine spacetime geometry. A general theory of this type is a metric-affine gravity[4] [10–13].

Remarkably, the coframe bundle has a contractible piece $GL(d,\mathbb{R})/SO(d-n,n)$ where $SO(d-n,n)$ is obviously a stability group. This means that the coframe bundle can be naturally contracted down to $SO_M = (SO(d-n,n), M)$ [14, 17, 19]. The fact that the contraction is topologically favored has drastic consequences to the geometry, it means that every manifold M can assume a Riemannian metric, *i.e.*, the connection can always be chosen to be compatible with the metric. This means that the metric tensor can be set as a constant flat one, $m = \eta$, where the signature of η depends on n. As a consequence, a standard fiber at X is the set of all orthonormal coframes that can be obtained from an $SO(d-n,n)$ transformation acting on a fixed coframe. The group $SO(d-n,n)$ describes then the isometries in $T^*_X(M)$. From Theorem 2.1 the connection $\Gamma = \omega + w$ imposes an $SO(d-n,n)$ connection $\omega \in \widetilde{O}$, where $\widetilde{O}$ is the algebra of $SO(d-n,n)$ and $w \in \widetilde{GL}/\widetilde{O}$. A gravity theory constructed over SO_M is a standard Einstein-Cartan gravity. In this work we shall deal strictly with SO_M.

To construct a moduli bundle for gravity is not immediate as in pure gauge theories. If the coframe bundle is a gauge bundle then e is actually a matter field because it is a fundamental representation of the gauge group [24]. On the other hand, one can include the space of all independent e that can be defined in $T^*_X(M)$ as the coframe moduli space $\mathcal{E}$. Thus, defining the full moduli space as $\mathcal{G} = \mathcal{W} \times \mathcal{E}$, where $\mathcal{W}$ is the moduli space of spin-connections, the

[4] Metric-Affine gravities can be also generalized for the affine group $A(d,\mathbb{R}) = GL(d,\mathbb{R}) \ltimes \mathbb{R}^d$, however, the non-semi-simplicity of this group spoils the construction of an invariant action. We shall fix our attention to semi-simple groups.

gauge orbit is then

$$\begin{aligned} \omega^g &= g^{-1}(\mathrm{d}+\omega)g\,, \\ e^g &= ge\,, \end{aligned} \tag{2}$$

with $g \in SO(d-n,n)$ and $W = (\omega, e) \in \mathcal{G}$. The moduli coframe bundle is $\mathbb{O} = (SO_M, \mathcal{G})$. This principal bundle is analogously equivalent to that described in Theorem 2.3. Thus, the space of all sections that can be defined over SO_M is actually the functional space of coframes. This space is equivalent to the fiber bundle $\mathbb{E} = (\Sigma(e), SO_M, \mathcal{E})$ where the fiber $\Sigma(e)$ is the set of all equivalent sections that can be obtain from an element $e \in \mathcal{E}$ through the action of SO_M.

4. Effective geometries

We now discuss the possibility of a gauge theory to be mapped into a gravity theory. We first discuss the map between gauge and coframe bundles and then we generalize the results for moduli bundles.

4.1 Gauge and coframe bundles

Theorem 4.1. *Let $H_R = (H, R)$ be a stable reduced bundle obtained from the gauge bundle $G_R = (G, R)$ which is endowed with a connection $Y = A + B$. Then G_R can define a geometry $SO_M = (SO(d-n,n), M)$ if and only if*

1. *The base spaces R and M are isomorphic;*
2. *The structure groups H and $SO(d-n,n)$ are related, at least, by a surjective homomorphism;*
3. *A composite field θ, which is an invariant representation of H, can be identified with an invariant fundamental representation of $SO(d-n,n)$.*

Proof. Condition *1* ensures that each point $x \in R$ can define a unique point in $X \in M$ while M will be entirely covered by the map with no overlapping points. Moreover, the algebraic structure defined in R will be preserved by the mapping. On the other hand, condition *2* ensures that the target group $SO(d-n,n)$ will be entire covered by the mapping. To construct the fiber at a cotangent space $T^*_X(M)$ in each point $X \in M$ we need two quantities: a coframe $e \in T^*_X(M)$ and the isometries of the cotangent space. The use of conditions *1* and *2* ensures the existence of the isometries. Since there is a fiber H in each point $x \in R$ and condition *2* ensures that H is at least homomorphic to $SO(d-n,n)$, this fiber defines the cotangent space $T^*_X(M)$ isometries. In addition, since there is one fiber for each point $x \in R$ there will be only one set of isometries for each $X \in M$, as it is evident from condition *1*. Condition *3* ensures that the field θ, in the fiber H_R, can be defined as the cotangent 1-form $e \in T^*_X(M)$, recognized as a coframe. Once more, the isomorphism of condition *1*, together with Corollary 2.4, ensures the uniqueness of e in X. Finally, a standard fiber in SO_R is obtained by the action of $SO(d-n,n)$ on e. A connection ω in SO_M emerges naturally from A. Again, condition *1* establishes that at a point X there will be only one ω while the action of H on A ensures that ω will transform correctly along the fiber $\pi^{-1}(X)$ under the action of the local isometries in $T^*_X(M)$. More explicitly, In each fiber $\pi^{-1}(x)$ a connection A can be defined. This definition ensures the existence of an equivalence class along the fiber. Thus, a section $s(x) : R \longmapsto H(x)$ is defined in such a way that $x \longmapsto q$, where $q = (x, g)$ and $g \in G$. In each point q the connection $A(q)$ can be identified with a connection $\omega(Q)$ in $SO(X)$ at a point $Q = (X, u) \in SO(X)$ and $u \in SO(d-n,n)$ is the $SO(d-n,n)$ equivalent of g such that $\pi(Q) = X$, $\pi(q) = x$ and $x \longmapsto X$. Condition *1* ensures that there will be only one connection $\omega(Q)$ for each $A(q)$.

Obviously, the reconstruction of the whole class of connections along a fiber is obtained from the action of the group on $\omega(Q)$. The final result is a mapping $G_R \longmapsto SO_R$ which is actually a set of mappings

$$\begin{aligned} R &\longmapsto M \,, \\ H &\longmapsto SO(d-n,n) \,, \\ \theta &\longmapsto e \,, \\ A &\longmapsto \omega \,. \end{aligned} \tag{3}$$

We remark that, in the mapping $G_R \longmapsto SO_R$, a contraction $G_R \longrightarrow H_R$ is assumed. □

Comment. If the map described in Theorem 4.1 is smooth and all fibers $\pi^{-1}(x)$ are mapped into fibers $\pi^{-1}(X)$ then this map is a bundle map. In that case, since each fiber $\pi^{-1}(x)$ is mapped into a fiber $\pi^{-1}(X)$ in a smooth way, a smooth map $R \longmapsto M$ is induced [20].

Comment. We remark that $\dim M = \dim T^*_X(M)$ and thus $\dim R = d_o$ is not necessary equal to $\dim M = d$. Notwithstanding, the bound $d \leq d_o$ is a always valid. Furthermore, d is the dimension of the fundamental representation of $SO(d-n,n)$, as a consequence it coincides with the dimension of the invariant representation of H, namely θ. The case $d < d_o$ affects only a subsector of spacetime $R \supset R_{sub} \longmapsto M$, where $\dim R_{sub} = \dim M$. For instance, if $R = \mathbb{R}^{d_o}$, then the resulting full manifold is then $M \times \mathbb{R}^{d_o - d}$. The case $d = d_o$ deforms the entire spacetime. This case is more interesting because one can take the starting gauge theory as a description for quantum gravity. from now on, independently of the case, we shall call by M the full d_o-dimensional manifold formed by the deformed (d-dimensional subspace) and undeformed (($d_o - d$)-dimensional subspace) sectors.

Corollary 4.2. *If the space of p-forms in R are directly mapped into the space of p-forms in M, $\Pi^p_R \longmapsto \Pi^p_M$, then the map can be explicitly computed and depends exclusively on the metric tensors of R and M.*

Proof. By duality the map $\Pi^p_R \longmapsto \Pi^p_M$ induces a similar map for the Hodge dual space of $(d-p)$-forms, $*\Pi^p_R \longmapsto \star\Pi^p_M$, where $*$ is the Hodge operation in R while $\star$ is the Hodge operation in M. Thus, it is a straightforward exercise [9] to show that the map is given by

$$\frac{\partial X^\nu}{\partial x^\mu} = \left(\frac{\tilde{m}}{m}\right)^{1/2d} \tilde{m}^{\nu\alpha} m_{\alpha\mu} \,, \tag{4}$$

where $m_{\mu\nu}$ is the metric tensor in R, $\tilde{m}_{\mu\nu}$ is the metric tensor in M and m and $\tilde{m}$ are the respective determinants. The determinants are assumed to be non-vanishing. □

Comment. Since the mapping matrix (4) has an inverse, the geometry in M is unique.

4.2 Moduli bundles and gravity

Theorem 4.1 can be generalized for moduli bundles:

Theorem 4.3. *If the map $G_R \longmapsto SO_M$ exists then the map $\mathbb{Y} \longmapsto \mathbb{O} \oplus \tilde{\mathbb{B}}$ also exists. The space $\tilde{\mathbb{B}}$ is the target space associated with the space $\mathbb{B}$ or $\mathbb{B}/\Theta$ if $\Theta \subseteq \mathbb{B}$ where Θ is the functional space of all possible θ that can be defined in $\mathbb{Y}$.*

Proof. In a principal principal bundle G_R a connection Y can be defined. The collection of all possible connections Y defines the space $\mathbb{Y}$. The separation of $\mathbb{Y}$ into equivalence classes organizes this space as the set of all gauge orbits over the moduli space $\mathcal{C}$. According to Theorem 2.3, the gauge orbit splits as (1). Moreover, for each of these fibers one can associate a field θ, as allowed by Corollary 2.4. It is the pair (A, θ) that defines the geometric fields (ω, e) in SO_M. Thus, to construct an $\mathbb{O}$ structure for gravity is an easy task to collect all possible pairs $W = (\omega, e)$ emerging from Theorem 4.1. In fact, each pair (A, θ) and the associated orbit define a fiber $W^g \in \mathbb{O}$. That is ensured also by Condition *1* of Theorem 4.1. Thus, the fiber H_R over a point $(\omega, e) \in \mathcal{G}$ is obtained from $A \longmapsto \omega$ and $\theta(A, B) \longmapsto e$ and the respective action of $SO(d-n, n)$. The uniqueness of this mapping is ensured by Corollary 2.4.

The space $\mathbb{B} = (\Sigma(B), H_R, \mathcal{B})$ is a dynamical space and survives the mapping. For the case $\mathbb{B} \cap \Theta = \emptyset$ one can associate the moduli space with a set of independent fields $\mathcal{B} \longmapsto \tilde{\mathcal{B}}$ which are invariant representations of $SO(d-n, n)$. Theorem 4.1 ensures that the structure group H_R can be mapped into SO_M while the fibers $\Sigma(B)$ are identified with fibers $\Sigma(\tilde{B})$ over $\tilde{B}$. Thus, for each $B \in \mathbb{B}$ there will be a correspondent $\tilde{B} \in \tilde{\mathcal{B}}$ and the fiber $\Sigma(\tilde{B})$ is obtained from the action of $SO(d-n, n)$. Thus, $\tilde{\mathbb{B}} = (\Sigma(\tilde{B}), SO_M, \tilde{B})$. The proof for the case $\mathbb{B} \cap \Theta \neq \emptyset$ is totally equivalent. □

Comment. The final result is that of a gravity theory with an extra set of matter fields $\tilde{\mathbb{B}}$.

5. Final remarks

We have formally prove that a class of gauge theories can be deformed into a first order gravity theory and, possibly, with an extra set of matter fields. For that we have employed the theory of fiber bundles. The relevance (and motivation) of the present work is that it can be applied to quantum gravity models which are based on gauge theories that can generate an emergent gravity theory. The main problem in quantizing gravity is that the principles of general relativity are incompatible with those of quantum field theory. In fact, a quantum field theory can only be formulated in an Euclidean spacetime. For example, a quantum field is, by definition, an object that carries uncertainty fluctuations and is parametrized through spacetime coordinates, *i.e.*, a set of well defined real parameters. Now, if a coframe field is a quantum field[5], $\hat{e}(x)$, and from the fact that it defines a mapping from tangent coordinates x^a to world coordinates x^μ, then quantum fluctuations of $\hat{e}$ will induce spacetime to fluctuate as well, $\hat{x}^\mu = \hat{e}^\mu_a x^a$. Thus, a paradox is encountered because x must be a set of parameters instead of a fluctuating object.

On the other hand, if the starting gauge theory is constructed over an Euclidean manifold and it is renormalizable, then it can be an excellent candidate for a quantum gravity theory. All needed is that it emerges as a geometrodynamics at classical level. The class of theories that fits on this program are determined essentially by theorems 4.1 and 4.3.

A few practical examples are in order: In [15] a 4-dimensional $SU(2)$ gauge theory generates a deformation of the 3-dimensional space. Time is left untouched by he mapping. In this example, the resulting theory contains the Einstein-Hilbert action for the extrinsic curvature and the solution of the equations of motion predicts not only curvature but also torsion. Another example can be found in [9], where a deformed 4-dimensional spacetime emerges from a de Sitter type gauge theory over an Euclidean spacetime. In this case, a dynamical mass scale is responsible for the separation between the gauge and gravity phases. In general,

[5] The hat indicates the quantum nature of the field.

several emergent gravity theories fit to the results of this work, see for instance [1–8] where the Higgs mechanism is largely used to separate gauge and gravity phases.

We end this work by remarking that several issues are left for future investigation. Just to name a few: The role of matter fields living in the starting gauge theory; the generalization of the present results to include metric-affine gravities before the reduction of the coframe bundle; the role of the extra matter fields in the dark matter/energy problem; explicit computations in order to make reliable predictions that fit with actual data; and so on.

6. Acknowledgements

Conselho Nacional de Desenvolvimento Científico e Tecnológico[6] (CNPq-Brazil) is acknowledge for financial support. I also would like to express my gratitude to InTech for this opportunity.

7. References

[1] S. W. MacDowell, F. Mansouri, Phys. Rev. Lett. 38, 739 (1977).
[2] K. S. Stelle, P. C. West, J. Phys. A A12, L205-L210 (1979).
[3] K. S. Stelle, P. C. West, Phys. Rev. D21, 1466 (1980).
[4] A. A. Tseytlin, Phys. Rev. D26, 3327 (1982).
[5] H. R. Pagels, Phys. Rev. D29, 1690 (1984).
[6] P. Mahato, Phys. Rev. D70, 124024 (2004).
[7] R. Tresguerres, Int. J. Geom. Meth. Mod. Phys. 5, 171-183 (2008).
[8] E. W. Mielke, Phys. Lett. B688, 273-277 (2010).
[9] R. F. Sobreiro, A. A. Tomaz and V. J. V. Otoya, arXiv:1109.0016 [hep-th].
[10] F. W. Hehl, G. D. Kerlick and P. Von Der Heyde, Phys. Lett. B 63, 446 (1976).
[11] F. W. Hehl, J. D. McCrea, E. W. Mielke and Y. Ne'eman, Phys. Rept. 258, 1 (1995) [arXiv:gr-qc/9402012].
[12] A. Trautman, Czech. J. Phys. B 29, 107 (1979).
[13] A. Trautman, "Fiber Bundles, Gauge Fields, And Gravitation," *In *Held.A.(Ed.): General Relativity and Gravitation, Vol.1*, 287-308*
[14] R. F. Sobreiro, V. J. Vasquez Otoya, J. Geom. Phys. 61, 137-150 (2011).
[15] Y. .N. Obukhov, Theor. Math. Phys. 117, 1308-1318 (1998).
[16] C. Itzykson and J. B. Zuber, "Quantum Field Theory," New York, Usa: Mcgraw-hill (1980) 705 P.(International Series In Pure and Applied Physics)
[17] S. Kobayashi and K. Nomizu, "Foundations of Differential Geometry, Vol. 1" *New York, USA: John Wiley & Sons (1963).*
[18] M. Daniel and C. M. Viallet, Rev. Mod. Phys. 52, 175 (1980).
[19] C. Nash and S. Sen, "Topology And Geometry For Physicists," *London, Uk: Academic (1983) 311p.*
[20] M. Nakahara, "Geometry, topology and physics," Bristol, UK: Hilger (1990) 505 p. (Graduate student series in physics).
[21] R. A. Bertlmann, "Anomalies in quantum field theory," Oxford, UK: Clarendon (1996) 566 p. (International series of monographs on physics: 91).
[22] R. F. Sobreiro and V. J. V. Otoya, J. Phys. Conf. Ser. 283, 012032 (2011) [arXiv:1010.2946 [hep-th]].
[23] B. McInnes, Class. Quant. Grav. 1, 1 (1984).
[24] V. A. Rubakov, "Classical theory of gauge fields," Princeton, USA: Univ. Pr. (2002) 444 p

[6] RFS is a level PQ-2 researcher under the program *Produtividade em Pesquisa*, 304924/2009-1.

6

Quantum Gravity in Cantorian Space-Time

L. Marek-Crnjac
Technical School Center of Maribor, Maribor
Slovenia

1. Introduction

Einstein's contribution to relativity was initially an intuitive approach based on a basic elimination of simultaneousity and a mathematical reformulation using the Lorenz transformation. In this respect Einstein just added some more physics to what Poincaré and Lorenz have done much earlier. However, it was Minkowski who introduced the geometrical ideas and the use of a four-dimensional space with time as the fourth dimension. Einstein took over Minkowski's idea and initiated what we may call the program of geometrizing physics, starting with gravity. Later on Einstein and Hilbert attempted the unification of electro-magnetism and gravity while Kaluza and Klein tried the same using an extra fifth dimension. This may have been the beginning of the higher dimensional space-time theories culminating in super strings, super gravity and the Cantorian space-time theory [1].

In special relativity there is no absolute time. We have a space and each slice has its own time. Thus each point in the Minkowski's space is specified by four coordinates, three spatial and one temporal in a four-dimensional space-time rather than the 3+1 space plus time coordinates of the classical mechanics [1].

A fundamental role in this new geometry is played by the constancy of the velocity of light that cannot be exceeded without violating the causal structure as well documented experimental facts show. A change of things began by adding quantum mechanics to special relativity.

By replacing Euclidean geometry by curved Riemannian one, Einstein was the first to give gravity a geometrical interpretation as a curvature of space-time due to matter. Einstein never fixed the topology of his theory nor did he use or was aware of the existence of non-classical geometry which was in any case in its infancy [1-5]. The possibly only encounter of Einstein with M. S. El Naschie's Cantorian like transfinite geometry was when K. Menger presented a paper in a conference held in his honour [1, 6-17].

2. Transfinite sets and quantum mechanics

Let us examine the basic concept of a line or more generally a curve. Classical geometry used in classical mechanics and general relativity the fact that a line is a one-dimensional object, while a point is zero-dimensional. Furthermore, it would seem at first sight that a line consists of infinite number of points and that it is simply the path drawn by a zero-dimensional point moving in the two or three-dimensional space. Classical geometry similar to classical mechanics has made various tacit simplifications and ignored several subtle topological facts [2].

If a line is one-dimensional and if it is made of infinite number of points then the sum of infinitely many zeros should be equal to one. That is of course not true. On the other hand we know that there is a curve called the Peano-Hilbert curve which is area filling and two-dimensional [1, 7-20]. By contrast, we can construct a three dimensional cube known as the Menger sponge which has a fractal dimension more than two and less than three, namely $D = \frac{\log 20}{\log 3}$ as explained for instance in the classical book of Mandelbrot [2].

The existence of all these non-conventional forms described in modern parlance, following Mandelbrot, as fractals, may be traced back to the archetypal transfinite set known as Cantor triadic set [6].

A Cantor set is a set of disjoint points which possesses the same cardinality as the continuum. It may be this coincidence that makes it an ideal compromise between the discrete and the continuum. It is transfinite discrete. Our Cantorian space-time which we will use to "topologize" physics is based on these transfinite sets. The main idea behind the Cantorian space-time approach is to replace the formal analysis of quantum mechanics and the Riemannian space-time geometry of general relativity by a transfinite fractal Cantorian space-time manifold [1, 8, 11, 13].

3. A short historical overview of ideas leading to fractal space-time

The idea of a hierarchy and fractal-like self-similarity in science started presumably first in cosmology before moving to the realm of quantum and particle physics [1]. It is possible that the English clergyman T. Right was the first to entertain such ideas (Fig. 1). Later on the idea reappeared in the work of the Swedish scientist Emanuel Swedenborg (1688-1772) and then much later and in a more mathematical fashion in the work of another Swedish astrophysicist C. Charlier (1862-1934) (Fig. 2).

In 1983, the English-Canadian physicist Garnet Ord wrote a seminal paper [3] and coined the phrase Fractal Space-time. Ord set on to take the mystery out of analytical continuation. We should recall that analytical continuation is what converts an ordinary diffusion equation into a Schrödinger equation and a telegraph equation into a Dirac equation. Analytical continuation is thus a short cut quantization. However what really happened is totally inexplicable. Ord showed using his own (invented) quantum calculus, that analytical continuation which consist of replacing ordinary time t by imaginary time it where $i = \sqrt{-1}$ is not needed if we work in a fractal-like setting, i.e. a fractal space-time. Although rather belated Ord's work has gained wider acceptance in the mean time and was published for instance, in Physics Review [4]. Therefore one is hopeful that his message has found wider understanding. It is the transfinite geometry and not quantization which produces the equations of quantum mechanics. Quantization is just a very convenient way to reach the same result fast, but understanding suffers in the process of a formal analytical continuation.

Fig. 1. A vision of T. Right's cosmos as a form of sphere packing, on all scales [1].

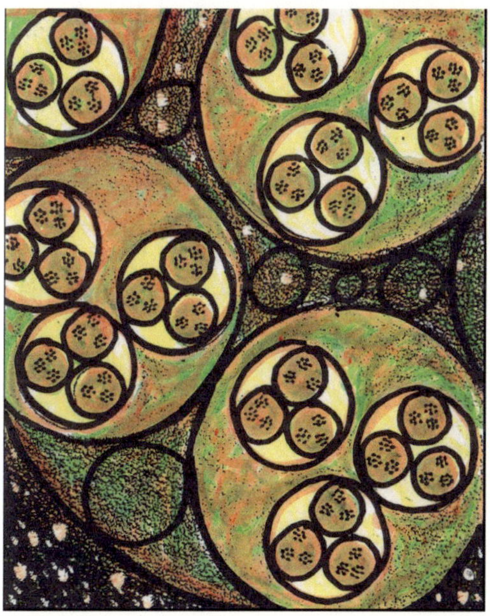

Fig. 2. A vision of a fractal-like universe, with clusters of clusters ad infinitum as envisaged by the Swedish astronomer C. Charlier who lived between 1862 and 1934. This work was clearly influenced by the work of the Swedish astrophysicist A. Swedenborg (1688–1772) [1].

Similar work, but not identical, was carried out by the French cosmologist Laurent Nottale, fifteen years ago. Nottale connected scaling and Einstein's relativity to what is now called scale relativity theory [5]. Around 1990, M. S. El Naschie began to work on his Cantorian version of fractal space-time [6]. In M. S. El Naschie's work on high energy physics and electromagnetic weak interactions the golden mean plays a very important role.

In the year 1995 Nobel laureate Prof. Ilya Prigogine, Otto Rössler and M. S. El Naschie edited an important book [7] in which the basic principles of fractal space-time were spelled out. Sometime later El Naschie using the work of Prigogine on irreversibility showed that the arrow of time may be explained in a fractal space-time. Recently El Naschie gave for the first time a geometrical explanation of quantum entanglement and calculated a probability of the golden mean to the power of five ϕ^5 for the entanglement of two quantum particles [8, 9].

4. Fractals

In this section, we give a very brief account of Cantor sets and fractals which are fundamental to the Cantorian space-time theory.

4.1 Triadic Cantor set and the random Cantor set

The archetypal fractal is what is known as Cantor triadic set. We start by describing the fundamental construction. Consider a unit interval. Let us delete the middle third but leave the end points. We repeat the procedure with the two segments left and so on as shown in Fig. 3 infinitely many times. At the end we obtain an uncountable set of points of measure zero. This means adding all these points together and we obtain a zero length. However, from the point of view of transfinite set theory something very profound is left, namely a transfinite points set with a finite dimension, the so-called Hausdorff-Besicovitch dimension [2]

$$d_C = \frac{\ln 2}{\ln 3} \cong 0.63$$

Mauldin and Williams replaced the orderly triadic construction by a random construction. In their original paper [10] they said they used a uniform probabilistic distribution. The Mauldin-Williams theorem which states that with the probability equal to one, a one dimensional randomly constructed Cantor set will have the Hausdorff-Besicovitch dimension

$$d_C^{(0)} = \frac{\sqrt{5}-1}{2}$$

The Menger-Urysohn dimension of all Cantor sets is zero, while the empty set has the dimension minus one [11].

4.2 The Sierpinski triangle, Menger sponge and their random analogues

The generalization of the one-dimensional triadic Cantor set to two-dimensions is called the Sierpinski triangle. It is constructed as shown in Fig. 3 and the Hausdorff-Besicovitch dimension is given by the inverse of the triadic Cantor set [2]

$$d_S = \frac{\ln 3}{\ln 2} \cong 1.5849$$

It is important to note that the Sierpinski triangle is a curve and its dimension lies between the classical line and the classical area.

Type of fractal	Geometrical shape	Menger-Urysohn dimension	Hausdorff dimension	Corresponding random Hausdorff dimension	Embedding dimension	Corresponding Euclidean shape
Cantor Set	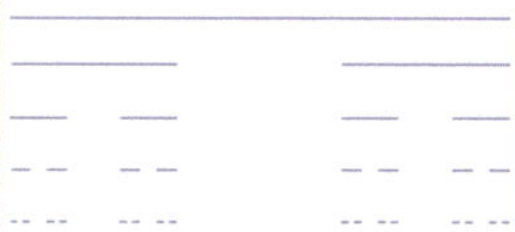	0	*d_C=ln2/ln3 0.630929*	$\phi = 0.618033$	1	*Line*
Sierpinski gasket		2	*d_S=ln3/ln2 1.5849625*	$\frac{1}{\phi} = 1 + \phi =$ 1.618033	2	 *Square*
Menger sponge		3	*d_M= ln20/ln3 2.7268*	$2 + \phi =$ 2.618033	3	 *Cube*
The 4 dimension random Cantor set		4	$d_C^{(4)} =$ *4.236068*	$4 + \phi^3 =$ 4.236067	5	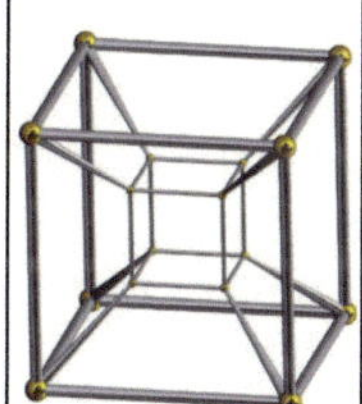 *Hyper cube*

Fig. 3. In this figure we draw analogy between smooth spaces as a line, a square, a cube, a higher-dimensional cube and the Cantor set, the Sierpinski triangle, the Menger sponge and the Cantorian space-time which is difficult to draw. The calculation of the Hausdorff-Besicovitch dimension of classical fractals and their random version is presented [1, 13].

It was shown in the Cantorian space-time theory [12] that the generalization of the formula connecting the triadic Cantor set with the Sierpinski triangle is possible for n dimension and is given by the so-called bijection formula

$$d_C^{(n)} = \left(\frac{1}{d_C^{(0)}} \right)^{n-1}$$

For $d_C^{(0)} = \phi$, the random contra part of the Sierpinski triangle will have the Hausdorff-Besicovitch dimension equal to [13]

$$d_C^{(2)} = \left(\frac{1}{\phi}\right)^{2-1} = \frac{1}{\phi} \cong 1.61803$$

A most remarkable 3D fractal is the Menger sponge which is shown in Fig. 3. The Hausdorff-Besicovitch dimension of this fractal is given by [2]

$$d_M = \frac{\ln 20}{\ln 3} \cong 2.7268$$

The volume of the Menger sponge is zero. The random version of the Menger sponge has a Hausdorff- Besicovitch dimension equal to [1, 13]

$$d_C^{(3)} = \left(\frac{1}{\phi}\right)^{3-1} = \left(\frac{1}{\phi}\right)^{2} = 2 + \phi$$

Using the bijection formula we can calculate any higher dimensional fractals [8, 11].

One of the most far reaching and fundamental discoveries using the zero measure Cantor sets is undoubtedly that of El Naschie probability of quantum entanglement. His result for two entangled particles is a generic and universal value of the golden mean to the power of five. This is exactly equal to the famous result of Lucien Hardy [8, 9]. Quantum entanglement is thus explained as a consequence of zero measure gravity. Similarly one could explain any velocity larger than the speed of light [8, 9].

5. Construction of a random Cantor set and the Cantorian space-time

The main idea of the Cantorian space-time theory is in fact a sweeping generalization of what Einstein did in his general relativity, namely introducing a new geometry of space-time which differs considerably from the space-time of our sensual experience. This space-time is taken to be Euclidean. By contrast, general relativity persuaded us that the Euclidean 3+1 dimensional space-time is only an approximation and that the true geometry of the universe in the large is in reality a four-dimensional curved manifold [1, 11].

In the Cantorian space-time theory we take a similar step and allege that space-time at quantum scales is far from being the smooth, flat and passive space which we use in the classical physics. On extremely small scales, at very high observational resolution equivalent to a very high energy, space-time resembles a vacuum fluctuation and in turn modeling this fluctuation using the mathematical tools of non-linear dynamics, complexity theory and chaos. In particular, the geometry of chaotic dynamics, namely the fractal geometry is reduced to its quintessence, i.e. Cantor sets. A Cantor set has no ordinary real physical existence, because its Lebegue measure is zero and nonetheless it exists indirectly because it does have a well defined non-zero quantity, namely its Hausdorff-Besicovitch dimension. The triadic Cantor set possesses a Hausdorff-Besicovitch dimension equal to $D = \frac{\log 2}{\log 3} \cong 0.63$.

For a randomly constructed Cantor set on the other hand, the Hausdorff-Besicovitch dimension is found to take the surprising value of the inverse of the golden mean $D = \frac{\sqrt{5}-1}{2} \cong 0.61803$ by virtue of the Mauldin-Williams theorem [10].

In 1986 R. Mauldin and S. Williams proved a remarkable theorem which confirmed the main conclusion of the Hausdorff-Besicovitch dimension of the Cantorian space-time. To explain the Mauldin-Williams theorem let us construct a Cantor set of the interval *[0, 1]* via a random algorithm as follows. First we chose at random an x according to the uniform distribution on *[0, 1]*, then between x and *1* we chose y at random according to the uniform distribution on *[x, 1]*. That way we obtain two intervals *[0, X]* and *[Y, 1]*. Next we repeat the same procedure on *[0, X]* and *[Y, 1]* independently and so on. Continuation of this procedure leads then to a random Cantor dust and the Hausdorff-Besicovitch dimension of this set will be with a probability one equal to $\phi = \frac{\sqrt{5}-1}{2} \cong 0.61803$ [10].

Cantorian space-time is made of an infinite number of intersections and unions of the randomly constructed Cantor sets. Let us denote the Hausdorff-Besicovitch dimension of these Cantor sets by $d_C^{(0)}$. Next we use $(d_C^{(0)})^n$ as a statistical weight for the topological dimension n=*1* to ∞ and determine the average dimension $\langle n \rangle$, i.e. the expectation value of n. This value is easy to find following the centre of gravity theorem of probability theory to be [1, 8, 11]

$$\langle n \rangle = \frac{\sum_{n=0}^{\infty} n^2 (d_C^{(0)})^n}{\sum_{n=0}^{\infty} n (d_C^{(0)})^n}$$

Since

$$\sum_{n=0}^{\infty} n (d_C^{(0)})^n = \frac{d_C^{(0)}}{(1-d_C^{(0)})^2}$$

and

$$\sum_{n=0}^{\infty} n^2 (d_C^{(0)})^n = \frac{d_C^{(0)}(1+d_C^{(0)})}{(1-d_C^{(0)})^3}$$

one finds that

$$\langle n \rangle = \frac{1+d_C^{(0)}}{1-d_C^{(0)}}.$$

Next let us calculate the average Hausdorff-Besicovitch dimension $\langle d_C \rangle$. We sum together all the Hausdorff-Besicovitch dimensions $(d_C^{(0)})^{(0)}$, $(d_C^{(0)})^{(1)}$, $(d_C^{(0)})^{(2)}$ ....., following the formula for the infinite convergent geometric sequence, $(d_C^{(0)})^{(0)}$, $(d_C^{(0)})^{(1)}$, $(d_C^{(0)})^{(2)}$,......, where $0\langle d_C^{(0)}\langle 1$, we obtain [1, 11]

$$\sum_{n=0}^{\infty}(d_C^{(0)})^n = \frac{1}{1-d_C^{(0)}}.$$

The average Hausdorff-Besicovitch dimension is thus

$$\langle d_C \rangle = \frac{\sum_{n=0}^{\infty}(d_C^{(0)})^n}{d_C^{(0)}} = \frac{1}{d_C^{(0)}(1-d_C^{(0)})}.$$

If the Cantorian space-time is to be without gapes and overlapping [1, 14] then we must set $\langle n \rangle$ equal to $\langle d_C \rangle$. Proceeding that way one finds from [1, 14] the following Peano-Hilbert space filling condition $\langle n \rangle = \langle d_C \rangle$ that

$$\frac{1+d_C^{(0)}}{1-d_C^{(0)}} = \frac{1}{d_C^{(0)}(1-d_C^{(0)})}.$$

Thus we have

$$(1+ d_C^{(0)}\,)\, d_C^{(0)} = 1$$

or

$$(d_C^{(0)})^{(2)} + d_C^{(0)} - 1 = 0.$$

This is a quadratic equation with two solutions

$$d_{C,1}^{(0)} = \frac{\sqrt{5}-1}{2} = \phi$$

$$d_{C,2}^{(0)} = -\frac{1}{\phi}.$$

Inserting back in $\langle n \rangle$ and $\langle d_C \rangle$ the solution $d_{C,1}^{(0)} = \phi$, one finds that

$$\langle n \rangle = \frac{1+\varphi}{1-\varphi} = \frac{1}{\varphi^3} = 4+\ \phi^3$$

and

$$\langle d_C \rangle = \frac{1}{\phi(1-\phi)} = \frac{1}{\phi^3} = 4 +\ \phi^3$$

where $\phi + \phi^2 = 1,\ \phi = \dfrac{\sqrt{5}-1}{2}.$

Our next aim is to solve the problem of lifting the random Cantor set $d_C^{(0)}$ to higher dimensions n and find $d_C^{(n)}$ for a given $d_C^{(0)}$.

The solution of this problem comes from the fact that the generalisation of the triadic set to two dimensions is the Sierpinski gasket. The Hausdorff-Besicovitch dimension of the gasket is the inverse value of the Hausdorff-Besicovitch dimension of the triadic set $\dfrac{\log 2}{\log 3}$.

Therefore one could write [1, 11]

$$d_C^{(2)} = \frac{1}{d_C^{(0)}} = \left(\frac{1}{d_C^{(0)}}\right)^{2-1}.$$

The generalisation by analogy and induction can thus be written as [12]

$$d_C^{(n)} = \left(\frac{1}{d_C^{(0)}}\right)^{n-1}.$$

Now let us examine the case for space filling, i.e. $d_C^{(0)} = \phi$ and four-dimensionality.

This way we obtain the Hausdorff-Besicovitch dimension of the Cantorian space-time

$$d_C^{(4)} = \left(\frac{1}{\phi}\right)^3 = 4 + \phi^3 \cong 4.236067.$$

This is a remarkable result which means that the formally infinite dimensional but hierarchical Cantorian space-time looks from a distance as if it were four-dimensional with the Hausdorff-Besicovitch dimension equal to $4 + \phi^3 \cong 4.236067$.

The preceding derivation could be regarded as a proof for the essential four-dimensionality of our physical space-time. We perceive space-time to be four-dimensional because this is the expectation value of our infinite dimensional Cantorian space-time.

6. Summing over paths and summing over all dimensions in the Cantorian space-time

We recall that Feynman gave an alternative formulation of quantum mechanics in which one calculates amplitudes by summing over all possible trajectories of a system weighted by $e^{\frac{is}{\hbar}}$, where s is the classical action, $i = \sqrt{-1}$ and $\hbar$ the Planck quantum. For one particle the path integral is thus [15]

$$Z = \int e^{\frac{is}{\hbar}} [dx]$$

where [dx] means that we are summing over all possible paths of the concerned particle. What is important here is to realize that from all of infinitely many paths which a quantum particle can take some are more probable than others. The probability of the actual path, that is to say the amplitude of an event is the sum over the amplitude corresponding to all paths. Thus we have a weight assigned to each path in the Feynman formulation of quantum mechanics.

In the Cantorian space-time theory we proceed in an analogous way. However, instead of summing over all paths, we sum over all dimensions of infinite dimensional hierarchical Cantorian space-time. El Naschie has recently demonstrated that E-Infinity is a Suslin operation and the so-called Suslin A operation [9]. In this theory Suslin scaling replace the classical Lagrangian and the classical calculus using descriptive set theory [16, 17].

7. Cantorian space-time and Newton's non-dimensional gravity constant

Quantum non-dimensional gravity constants can be derived from descriptive set theory [16]. In descriptive set theory and theory of polish spaces it is shown that [16, 17]:

Definition 1:

When a space A^N is viewed as the product of infinitely many copies of A with discrete topology and is completely metrizable and if A is countable then the space is said to be polish.

Two cases are of considerable importance.

Definition 2:

When a space is polish and when $A = 2 = [0,1]$, then we call $C= 2^N$ the Cantor space.

Definition 3:

When a space is polish and when $A = N$ then we call $B = N^N$ the Baire space.

Now we can proceed to explain the relationship between the Cantor space and Cantorian space-time. The relationship comes from the solution of the cardinality problem of a Borel set in polish spaces. Thus, we call a subset of a topological space a Cantor set if it is homeomorphic to a Cantor space [16, 17].

Theorem 1:

Let X be polish and $Y \subseteq X$ be a Borel set. Then either Y is countable or else it contains a Cantor set. In particular every uncountable standard Borel space has cardinality 2.

A Cantor space is homeomorphic to a triadic Cantor set and also to the random Cantor set [17]. The relation between the triadic Cantor set and the Cantor space establishes the relationship between the Cantor space and the Cantorian space-time, since the Cantorian space-time is a hierarchical infinite dimensional Cantor set with the expectation Hausdorff-Besicovitch dimension $4+\phi^3 \cong 4.236067$.

In particular, it has been shown [17] that when interpreting $\frac{1}{d_C^{(0)}}$ in the bijection formula as the average $\left\langle \frac{1}{d_C^{(0)}} \right\rangle = 2$ of the fundamental Wisse-Abbot theorem and taking N = 128,

($\overline{\alpha}_{ew} = 128$ is the inverse coupling constant measured at the electroweak scale) then the bijection formula

$$d_C^{(n)} = \left(\frac{1}{d_C^{(0)}}\right)^{n-1}$$

gives for $N = n = \overline{\alpha}_{ew} = 128$ the following

$$d_C^{(128)} = \left(\left\langle\frac{1}{d_C^{(0)}}\right\rangle\right)^{128-1} = 2^{127} \cong (1.70141)(10)^{38}$$

where $C = 2^N = d_C^{(128)}$. The value $(1.70141)(10)^{38}$ is the non-dimensional gravity constant $\overline{\alpha_G}$ which is defined as

$$\overline{\alpha_G} = \frac{\hbar c}{G m_p^2} \cong (1.7)(10)^{38}$$

It is of interest to mention that a similar result was found empirically by F. Parker Rhodes which was the subject of extensive discussions by Noyes [18]

$$\overline{\alpha_G} = 2^{127} + 137 = (1.7)(10)^{38}$$

Here $\hbar$ is the Planck quantum, c the speed of light, G the Newton's gravity constant, m_p the Planck mass.

8. Cantorian space-time and the connectivity dimension

Next we show the logarithmic scaling which will connect the non-dimensional gravity constant to the most fundamental equation namely the bijection formula. We start by taking the logarithm of both sides of the equation

$$\ln d_C^{(n)} = \ln\left(\frac{1}{d_C^{(0)}}\right)^{n-1}.$$

That means

$$\ln d_C^{(n)} = (n-1)\ln\left(\frac{1}{d_C^{(0)}}\right)$$

solving for n one finds that

$$n = \frac{\ln d_C^{(n)}}{\ln\left(\frac{1}{d_C^{(0)}}\right)} + 1.$$

Setting $\left(\frac{1}{d_C^{(0)}}\right) = 2$ and $d_C^{(n)}$ =Z, where Z is the partition function, one finds

$$n = \frac{\ln Z}{\ln 2} + 1.$$

The above formula is very well-known in the combinatorial topology [14, 19, 20] and is called the connectivity dimension. Now if we conceive of $\overline{\alpha_G}$ as being the expectation value of the partition function of the observable universe then the connectivity dimension would be

$$D = \frac{\ln \overline{\alpha_G}}{\ln 2} + 1 \cong 128 = \overline{\alpha_{ew}}$$

This is the inverse of the Sommerfeld electromagnetic fine structure constant measured at the electroweak scale [1, 13].

9. Fundamental constants of Cantorian space-time

The fine-structure constant usually denoted with *a*, is a fundamental physical constant, namely the coupling constant characterizing the strength of the electromagnetic interaction. It is a dimensionless quantity and is defined as

$$\alpha = \frac{e^2}{(4\pi\varepsilon_0)\hbar c} = \frac{1}{137.035999074}$$

or as the inverse fine-structure constant

$$\overline{\alpha} = \frac{1}{\alpha} = 137.03599907$$

where e is the unit electron, $\hbar = h/2\pi$ is the Planck constant, c is the speed of light, ε_0 permittivity of free space.

In the Cantorian space-time theory the inverse fine-structure constant $\overline{\alpha_0}$ can be written in a remarkable short form based upon the multiplication and addition theorems of probability theory [1]. This is done by interpreting $d_C^{(0)} = \phi$ as a topological probability of a Cantor set formed by the ratio of the Hausdorff- Besicovitch dimension $d_C^{(0)} = \phi$ and the embedding topological dimension $d_C^{(1)} = 1$.

That way one finds

$$\overline{\alpha_0} = (2)(10)\left(\frac{1}{d_C^{(0)}}\right)^4$$

or

$$\overline{\alpha_0} = (2)(10)\left(\frac{1}{\phi}\right)^4 = 137.082039.$$

The value 137.082039 is in excellent agreement with the measured experimental value.

From the inverse fine-structure constant $\overline{\alpha_0}$ we can derive the inverse coupling constant of the non-super symmetric unification of all forces $\overline{\alpha_g}$ and the inverse coupling constant of the super symmetric unification of all forces $\overline{\alpha_{gs}}$ using the scaling arguments in the Cantorian space-time [1] . The scaling factor in the Cantorian space-time is ϕ . To derive the inverse coupling constant of the non-super symmetric unification of all forces $\overline{\alpha_g}$ we start with the Cooper pair. That means we multiply $\frac{\overline{\alpha_0}}{2}$ with ϕ and obtain the following result [20]

$$\frac{\overline{\alpha_0}}{2}\phi = 42.360679 = \overline{\alpha_g}.$$

Proceeding in this way one finds the inverse coupling constant of the super symmetric unification of all forces $\overline{\alpha_{gs}}$. We multiply $\frac{\overline{\alpha_0}}{2}$ with ϕ^2 and obtain

$$\frac{\overline{\alpha_0}}{2}\phi^2 = 26.18033989 = \overline{\alpha_{gs}}.$$

Both inverse coupling constants are in full agreement with the experimental values [1, 13, 20].

10. Conclusion

In the present review article we gave a short overview of ideas leading to the fractal space-time and the Cantorian space-time theory. The triadic set, the Sierpinski gasket, the Menger sponge and their random analogous are introduced. The Cantorian space-time is determined by three dimensions, the formal $n_f = \infty$, the topological n_T = 4 and the Hausdorff-Besicovitch dimension equal to $4 + \phi^3 \cong 4.236067$.

Feynman introduced a procedure which consists of summing over all possible paths of the concerned particle. In the Cantorian space-time theory the procedure is analogous, but instead of summing over all paths we sum over all dimensions of the infinite dimensional but hierarchical Cantorian space-time.

We establish a conceptual and quantitative connection between classical gravity and the electro-weak field using the Cantorian space-time theory and the descriptive set theory. This led El Naschie to a fundamental discovery for quantum entanglement [8, 9, 21, 22].

With the use of the golden mean scaling operator we derive an expectation value of the inverse electromagnetic fine structure constant $\overline{\alpha_0}$, the inverse coupling constant of the non-super symmetric unification of all forces $\overline{\alpha_g}$ and the inverse coupling constant of the super symmetric unification of all forces $\overline{\alpha_{gs}}$.

11. Acknowledgement

The author would like to thank Prof. Mohamed El Naschie for the discussion and permission to use his figures.

12. References

[1] El Naschie M. S.: A review of E-infinity theory and the mass spectrum of high energy particle physics. Chaos, Solitons and Fractals 19, (2004); 209-236.

[2] Mandelbrot B.:The fractal geometry of nature. San Francisco: Freeman; 1982.

[3] Ord G.: Fractal Space-time. J. Phys. A: Math. Gen. 16, 1869 (1983).

[4] Ord G.: Entwined paths, difference equations and the Dirac equation. Physics Review A (2003), 67: 0121XX3.

[5] Nottale L.: Fractal Space-time and microphysics. Singapore: World Scientific (1993).

[6] El Naschie M. S.: On the Universal Behavior and statistical Mechanics of Multidimensional Triadic Cantor Sets. SAMS, 1993, vol 11, 217-225.

[7] Prigogine I., Rössler O. and El Naschie M.S.: Quantum mechanics, diffusion and chaotic fractals. Pergamon (Elsevier-ISBN 0 08 04227 3) (1995).

[8] El Naschie M. S.: Quantum Collapse of Wave Interference Pattern in the Two-Slit Experiment: A set Theoretical Resolution. Nonlinear Science Letters A, Vol. 2, No. 1, 2011, pp. 1-9.

[9] He J. H. et al.: Quantum Golden Mean Entanglement Test as the Signature of the Fractality of Micro Space-time. Nonlinear Science Letters B, Vol. 1, No. 2, 2011, pp. 45-50.

[10] Mauldin R. D., Williams S. C.: Random recursive construction. Trans. Am. Math. Soc. 295: 325-346, (1986).

[11] Marek-Crnjac L.: The Hausdorff Dimension of the Penrose Universe. Physics Research International, Vol. (2011), 2011.

[12] El Naschie M. S.: Superstrings, knots and non-commutative geometry in E- infinity space. Int. J. Theoret. Phys. 1998; 37 (12).

[13] El Naschie M. S.: Elementary prerequisites for E-infinity (Recommended background readings in nonlinear dynamics, geometry and topology). Chaos, Solitons and Fractals, 30 (2006), 579-605.

[14] El Naschie M. S.: The concept of E-Infinity: An elementary introduction to the Cantorian–fractal theory of quantum physics. Chaos, Solitons and Fractals, 22 (2004), 495-511.

[15] Marek Crnjac L.: A Feynman path like-integral method for deriving the four dimensionality of space-time from first principle. Chaos, Solitons and Fractals, 41, (2009); 2471-2473.

[16] Jech T.: Set theory. Berlin: Springer; (2003).

[17] El Naschie M. S.: Quantum gravity from descriptive set theory. Chaos, Solitons and Fractals, 19 (2004), 1339-1344.

[18] Noyes H. P.:Bit-string physics. Singapore: World Scientific; (2001).

[19] Alexander P. S.: Combinatorial topology. Dover: (1998).

[20] Marek Crnjac L.: On the unification of all fundamental forces in a fundamentally fuzzy Cantorian $E^{(\infty)}$ manifold and high energy particle physics. Chaos, Solitons and Fractals, 20 (2004), 669-682.

[21] El Naschie M. S.: Quantum entanglement as a consequence of a Cantorian micro space-time geometry. J. Quantum Info. Sci.,1, (2011), p. 50-53.

[22] Marek-Crnjac L.: The common sense of the geometry of quantum entanglement. Fract. Spacetime Noncommut. Geom. Quant. High Energ. Phys., Vol.1, No.1, (2011), p. 31-39.